AROMA RESEARCH

Proceedings of the International Symposium on Aroma Research held at the Central Institute for Nutrition and Food Research TNO, Zeist, the Netherlands, May 26–29, 1975

Editors: H. Maarse and P.J. Groenen

Wageningen
Centre for Agricultural Publishing and Documentation
1975

ISBN 90 220 0573 9

© Centre for Agricultural Publishing and Documentation, Wageningen, 1975

No part of this book may be reproduced and published in any form, by print, photoprint, microfilm or any other means without written permission from the publishers

Cover design: Pudoc, Wageningen

Printed in the Netherlands

D
664.07
INT

This book is dedicated to Dr C. Weurman, who instigated this symposium

Dr C. Weurman (1912 – 1975)

Contents

Preface

This book contains the papers and short contributions presented at the Symposium on Aroma Research, which was held at the Central Institute for Nutrition and Food Research TNO, Zeist, the Netherlands, May 1975. The symposium was organized on the instigation of the late Dr C. Weurman, to whom we dedicate the Proceedings.

An objective of this symposium was to provide ample opportunity for thorough discussion and informal contact within a relatively small group of participants (about 50). To keep the group small, the symposium was not announced, all participants being personally invited. This, of course, raised the problem of making choices.

On the last day of the meeting a large majority of the participants agreed that there is a need for this type of small informal symposium on aroma research and that such meetings should follow, at intervals of three years, in different countries. There was also almost unanimous agreement to name these symposia the Weurman Flavour Symposia. The first symposium was sponsored by the Organization for Nutrition and Food Research TNO, to which we express our gratitude. We gratefully acknowledge the help of the staff of Pudoc who did most of the editorial work.

H. Maarse
P. J. Groenen

Opening address

J. van Mameren

Organization for Nutrition and Food Research TNO, The Hague, the Netherlands

Ladies and Gentlemen,

First of all, I extend a cordial welcome to all of you who are participants in this symposium. In doing so, it is my pleasure to address you on behalf of the Netherlands Organization for Applied Scientific Research, which, as you know, is briefly called TNO.

Some time ago, the late Dr Weurman suggested the initiative to convene the present symposium and I immediately approved his idea. And that for three reasons. One: I knew Weurman's recognized expertise in aroma research. Two: I knew he could organize a symposium adequately. Three: I then held and still hold the view that a small informal gathering of a few international specialists is a suitable medium for discussions on joint problems.

You know that Weurman's death has meant a considerable loss to TNO and to our Central Institute for Nutrition and Food Research. Not only was he one of our best research workers; he was a good colleague and friend.

Dr Weurman started our aroma research in 1956. In those early days, he had many contacts with colleagues abroad; in particular the United States, where he worked for about a year and a half and where he conducted research on various topics including enzymic formation of aromatics in raspberries.

In the 1960s, Weurman's department grew rather rapidly; particularly because of added interest from industry. Weurman devoted much attention to the development of those analytical techniques where the aroma composition was altered as little as possible. His work was also recognized internationally, as was evidenced for example at an American Chemical Society meeting at Chicago in 1967.

I suggest, ladies and gentlemen, that, in the present symposium, we all try to conduct the discussions in the atmosphere of true friendship and good-fellowship which the late Dr Weurman always aimed at.

TNO is honoured to receive you experts on aroma research. We are convinced that this symposium will create new opportunities and stimulate continued research. You will appreciate that this small country, the Netherlands, attaches a high value to international co-operation.

At the moment, TNO employs some 4 500 people. About 10% of them work in the institutes of our Organization for Nutrition and Food Research and some 300 of those 450 people work here, in our Central Institute for Nutrition and Food Research. In fact, this institute is the largest – in terms of staff – of the entire TNO Organization. This fact, as such, need not be important. However, thanks to this institute's several departments, some discipline-oriented, others product-oriented, the aggregate of research facilities in food and nutrition research is certainly considerable. Government agencies as well as industry and commerce can satisfactorily tap the resources of this institute.

From all this, you might conclude that my task, as president of TNO's Organization for Nutrition and Food Research and as acting vice-president of TNO's Central Organization, is honorable and pleasant. That is, of course, to some extent true. The difficulty is that there is a fair amount of criticism of fund-devouring aspects of research and development activities in the Netherlands as elsewhere. Furthermore, here, too, doubts are expressed as to the benefit and usefulness of R & D. The upshot is that reorganizations are being considered that should enable the governmental authorities to get a better grip on scientific institutes and on research programmes. Obviously – and that is another cup of tea – we at TNO are by no means immune to continued financial inflation. Hence, as you will realize, the very funding of a research organization is a matter of growing concern. One might even fear that, in the remote future, problems of investment could arise.

I should add, in this context, that the research programme of our Organization for Nutrition and Food Research has so far met with little criticism, generally speaking.

As in other countries, large sections of the Dutch population express anxiety about the quality of our foodstuffs. There are people who fear that, slowly but steadily, the Dutch nation is being poisoned by all sorts of chemicals in food, either as residues or as additives. A common view too is that all modern food technology achieves is a multitude of products that are significantly worse in taste, aroma and consistency than the products of the past.

Well, it may be of interest for you to know that our Organization for Nutrition and Food Research has a budget of 27 million Dutch guilders. Now, more than half that sizable sum is being spent on research on quality of foodstuffs and agricultural products. Moreover, much research capacity is being used for biological and toxicological research.

My conclusion – and I hope you share it – is that we are doing useful work.

Assuming that the taste and aroma of man's food should be improved here and there – which is a point that I personally do not doubt – I feel certain that for you, as experts in those fields, to-day's world, and to-morrow's, is bound to leave scope for your continued research.

It is my hope that this symposium will help you to specify research needs, and relevant action, in your own fields of specialization. On behalf of our TNO Organization, I wish you a pleasant stay in the Netherlands, and, of course, I hope that your discussions here, at our Central Institute for Nutrition and Food Research, will be satisfactory in all respects.

Thank you.

26 May – Formation of aroma compounds in foods

Proc. int. Symp. Aroma Research, Zeist, 1975. Pudoc, Wageningen.

Biochemical formation of aroma components

F. Drawert

Institut für Chemisch-Technische Analyse und Chemische Lebensmitteltechnologie der Technischen Universität München, 8050 Freising-Weihenstephan, Bundesrepublik Deutschland

Abstract

The question is critically discussed what final level of knowledge has been reached after a period of about 25 years of inventory of aroma components. We have to distinguish between primary (genuine) aroma components, for example as we find them with... the intact cell structure of fruits, and secondary aroma components which may be formed as a result of technological processes and which are formed when cell structures of fruits are destroyed. In this connection we examplify, especially, enzymic processes.

Furthermore, the results of investigations on the biosynthesis of bitter substances of hops are discussed. The pathway of synthesis of the monoterpenes and sesquiterpenes such as humulone and lupulone are investigated and schematized being derived mainly from carboxylic acids. This is followed by an explanation of what we termed the 'biological' resp. the 'biotechnical sequences'. These 'sequences' allow, for example, an explanation of the relationship between substrate composition and the fermentation.

Finally there follows an exemplary discussion of the question to what extent biochemical knowledge can be used today for a biotechnological-biosynthetical production of aromas.

We have historical experiences with aromas. We know that there has been no high form of human culture without its special aromas, for instance in aroma-rich foods and beverages. Fruit-bearing trees, bushes or plants have shared with man in an evolution, for instance when fruit have been selected not only for size, colour or texture, but especially for their aroma, a heritage that threatens to disappear gradually. A. H. Ruys, a man who had lifelong experience of aromas, called aroma the fruit's soul. In 1961 he wrote, "Si aujourd'hui, je vous dis que c'est l'âme du fruit, c'est que l'arôme nous offre l'expression la plus parfaite de la réalité du fruit. L'arôme du jus d'un fruit, bien plus que son image visuelle, nous apprend de ce fruit tout ce que nous voulons savoir." (Ruys, 1962).

Aroma is determined by habit and by region of the world. It consists of many components graded in concentration and physiological effect. One could perhaps now say that certain aroma substances are psychopharmaceuticals. People have attempted to explain odours by vibrational and stereochemical theories (Merkel, 1972; Hail, 1963; Ohloff & Thomas, 1971; Moncrieff, 1951; Amoore, 1962, 1963, 1967). If aromas are classed into odours and tastes as is common practice in science, it can be calculated that there are probably more than 10^6 possible sensations of odour and only a few, perhaps 6, sensations of taste. There are plausible theories about how we sense the tastes sweet and bitter (Shallenberger, 1971; Belitz, 1973).

If we take account and seriously ask what we know about aromas, we must honestly answer "Not very much." This may be mainly, as we at least know, because most natural aromas consist of hundreds of components. The specific notes of these aromas derive from only a few substances of high stereospecificity and sometimes of extremely low threshold concentrations. The development of gas chromatography, especially gas distribution chromatography, the development of separation columns with several hundred thousand possible theoretical plates, and the development of detectors sensitive to 10^{-12} $g \cdot s^{-1}$ introduced an era of stock-taking, bringing a flood of publications on aroma substances. Luckily we have lists, such as 'Lists of volatile compounds in foods' from the Central Institute for Nutrition and Food Research TNO at Zeist in the Netherlands. There are also trials at computer storage of data. But what should be put on the computer? Perhaps the amounts detected in gaschromatographic peaks in relation to thresholds, odour qualities and empirical composition, perhaps to attain 'identical aromas'. Undoubtedly such a development of aroma profiles is under way with the work of Jennings, Wick, Teranishi, Ohloff and others. These profiles appear 'three-dimensional', with sensory peaks behind the gas chromatographic peaks giving an aroma picture.

Like many others, we have contributed more or less thoughtlessly to the catalogue of components detected by gas chromatography, in particular aromas of plant origin. But since our interest lay in the biosynthesis and biological origin of aroma substances, we had to fix conditions in the cell to a certain moment of time, for biochemical reasons. In other words, we had to disturb the cell structure in such a way that at the moment of deterioration, enzymic processes were inhibited. So we soon noticed, for instance, that at the moment fruit are homogenized, certain

Table 1. Formation of C_6 aldehydes and C_6 alcohols (μg/100 g) in grapes and in vine leafs.

1 = Grapes (Sbl. 2-19-58 – newly cultivated vine-variety: European x American). 2 = Vine leafs of the same variety. a = Instant inhibition of the enzymes. b = Inhibition of the enzymes after 10 min.

	1		2	
	a	b	a	b
Hexanal	15	300	90	410
Hex-*cis*-3-enal	10	60	0	300
Hex-*trans*-3-enal	2	10	0	100
Hex-*trans*-2-enal	10	1700	130	13600
+[1]	5	20	10	850
1-Hexanol	0	160	5	30
Hex-*cis*-3-en-1-ol	5	500	150	2400
Hex-*trans*-2-en-1-ol	3	170	170	260

1. + = Not identified for certain.

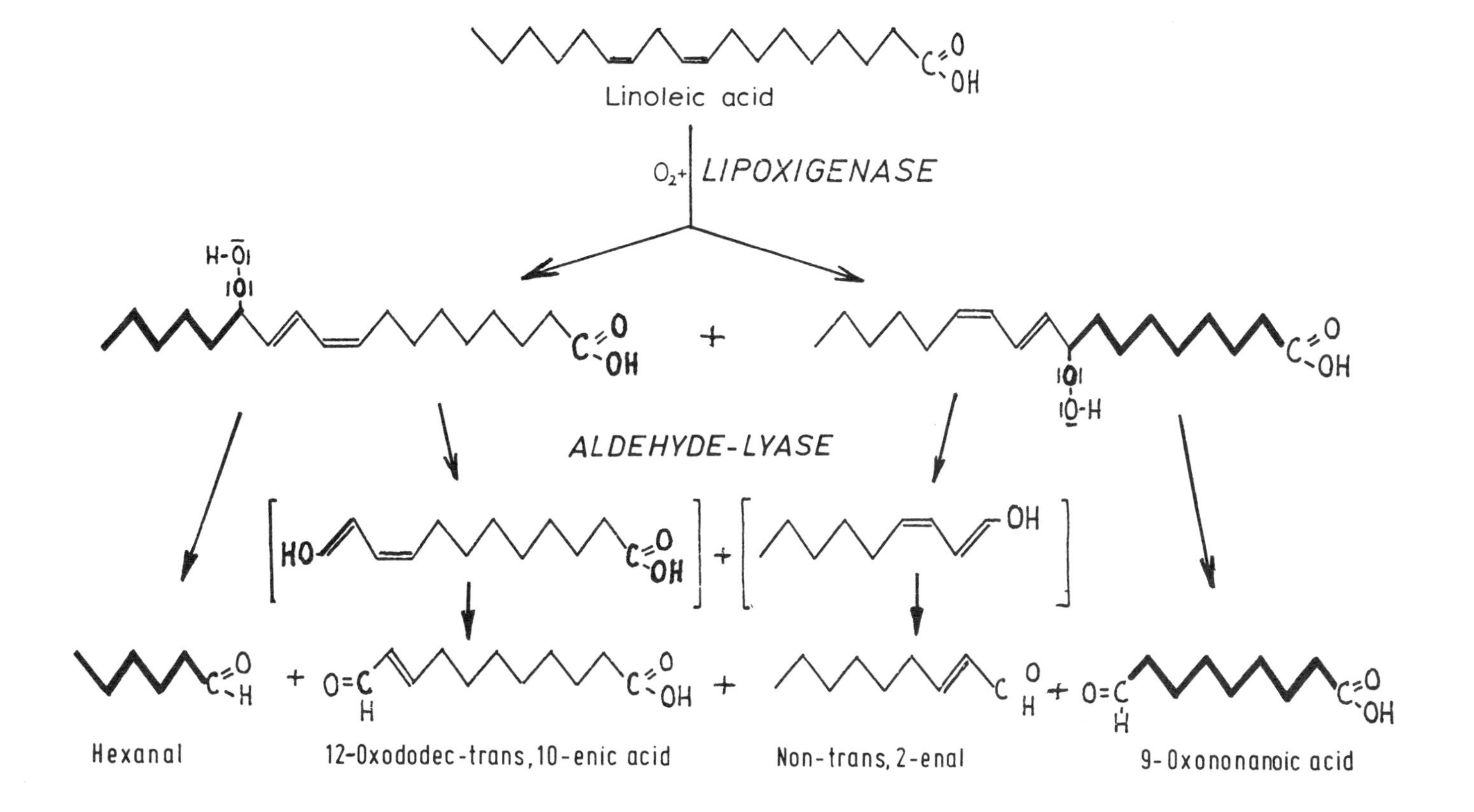

Fig. 1. Enzymic oxidative splitting of linoleic acid.

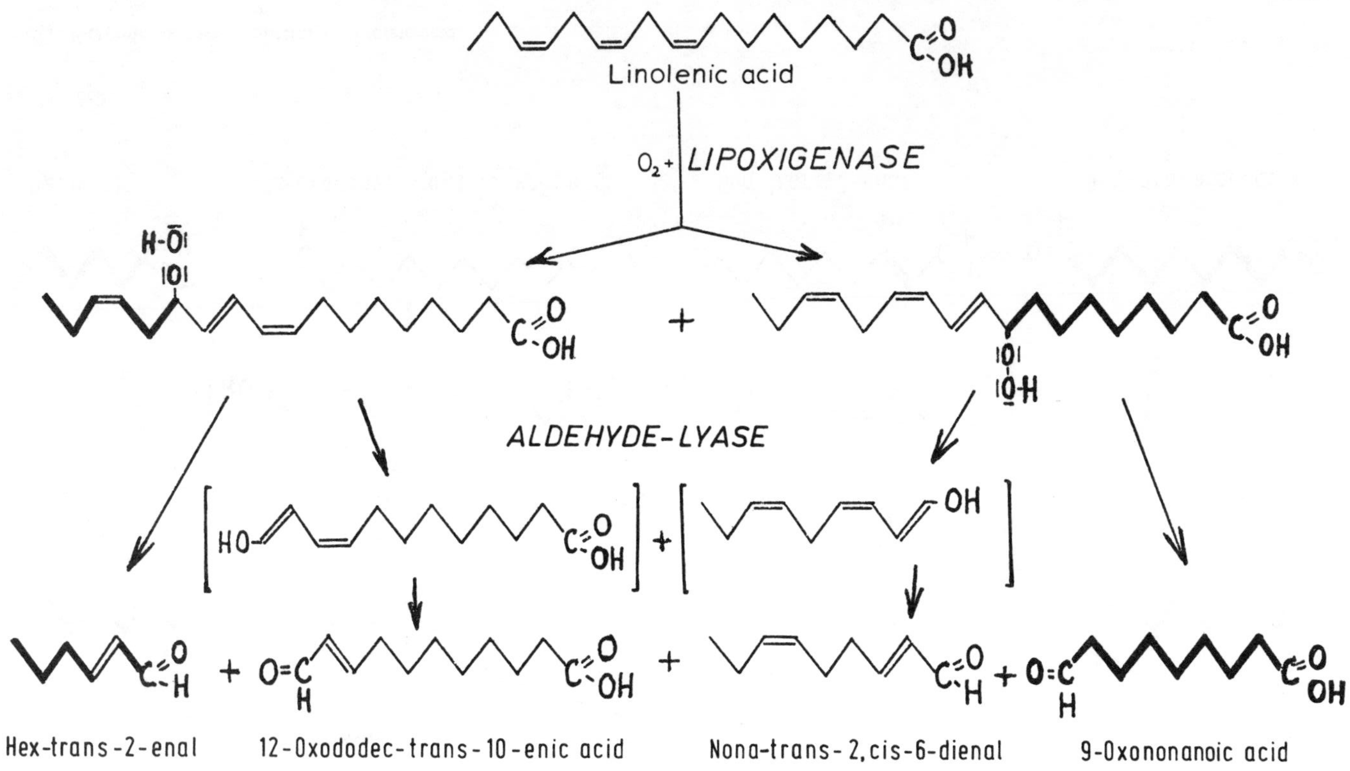

Fig. 2. Enzymic oxidative splitting of linolenic acid.

enzymic processes begin, some of them extremely rapid. In this way, for instance, esters characteristic of a fruit are hydrolyzed under the influence of hydrolases more or less rapidly according to pH and temperature (Drawert et al., 1965). In particular, enzymic oxidations set in and, as it were, contribute to the formation of new aroma substances (Drawert et al., 1966, 1969). Because of this, we learnt to distinguish strictly between primary or original aroma substances, and secondary or technological aroma substances.

In a short time, significant amounts of hexenals and hexanals arise (Table 1), whereby the various hexenals are specifically formed by the different fruit. Green bananas and cucumbers form *trans*-2-nonenal and *trans*-2,*cis*-6-nonadienal. Partly from studies with ^{14}C-labelled compounds, we concluded that these substances were fragments from linoleate and linolenate (Fig. 1 and 2; Drawert & Tressl, 1970; Tressl & Drawert, 1973).

Table 2. Hydrocarbons after irradiation, heating and smoking of lard.

Irradiation 0.5 – 6 Mrad		Heating		Smoking (outer layers)	
1.	Octane	1.	Octane		–
2.	Octene		–		–
3.	Nonane	2.	Nonane	1.	Nonane
4.	Nonene	3.	Nonene	2.	Nonene
5.	Decane	4.	Decane	3.	Decane
6.	Decene	5.	Decene	4.	Decene
7.	Undecane (4 isomers)	6.	Undecane	5.	Undecane
8.	Undecene	7.	Undecene	6.	Undecene
9.	Dodecane	8.	Dodecane	7.	Dodecane
10.	Dodecene (2 isomers)	9.	Dodecene (2 isomers)	8.	Dodecene
11.	Tridecane	10.	Tridecane	9.	Tridecane
12.	Tridecene (2 isomers)	11.	Tridecene (2 isomers)	10.	Tridecene
13.	Tetradecane	12.	Tetradecane	11.	Tetradecane
14.	Tetradecene (2 isomers)	13.	Tetradecene (2 isomers)	12.	Tetradecene
15.	Tetradecadiene		–		–
16.	Pentadecane	14.	Pentadecane	13.	Pentadecane
17.	Pentadecene	15.	Pentadecene (2 isomers)	14.	Pentadecene
18.	Pentadecadiene		–		–
19.	Hexadecane	16.	Hexadecane	15.	Hexadecane
20.	Hexadecene	17.	Hexadecene (2 isomers)	16.	Hexadecene
21.	Hexadecadiene		–		–
22.	Hexadecatriene		–		–
23.	Heptadecane	18.	Heptadecane	17.	Heptadecane
24.	Heptadecene (2 isomers)	19.	Heptadecene	18.	Heptadecene
25.	Heptadecadiene	20.	Heptadecadiene		–
26.	Heptadecatriene		–		–
27.	Octadecane	21.	Octadecane	19.	Octadecane
28.	Octadecene	22.	Octadecene	20.	Octadecene
		23.	*Ethylcyclohexene*		
		24.	*Propylcyclohexene*		
		25.	*Butylcyclohexene*		
		26.	*Pentylcyclohexene*		
		27.	*Hexylcyclohexene*		
		28.	*Heptylcyclohexene*		

Table 3. Formation of volatile compounds after irradiation with 6 Mrad.
Identification: gas chromatography – mass spectrometry.

Irradiation of:	Alcohols	Aldehydes	Ketones
Lard	1-pentanol 1-hexanol oct-1-en-3-ol 1-heptanol 1-octanol	butanal pentanal hexanal heptanal hex-2-enal octanal hept-2-enal nonanal oct-2-enal	2-heptanone 2-octanone 2-nonanone 2-decanone 9-oxononanoic acid
Linolenic acid methyl ester			9-oxononanoic acid[1] 12-oxododecanoic acid[1]
Linoleic acid methyl ester			9-oxononanoic acid[1] 12-oxododecanoic acid[1]
Oleic acid methyl ester		nonanal	9-oxononanoic acid[1]
Sunflower seed oil			9-oxononanoic acid

1. Methyl ester.

The fragments could also be recovered as the aldehydic carboxylates 12-oxo (*trans*-10)dodecenoic acid and 9-oxononanate. These are typical of the formation of secondary aroma substances, which according to physiological principles are formed partly as protective agents and partly as wound hormones. Even at minute concentrations, they give strong odours and tastes. The well known smell and taste of grass are actually due to hexanal and hexenals. Some authors have found these substances in apple juice and consider them characteristic, though they do not occur in intact apples. Even if that conclusion is false, it goes to show a certain habit or, in a legal sense, a certain expectation by the consumer, in so far as the consumer is accustomed to the factory product and to its technological aroma. Perhaps he even demands it.

Noteworthy is that radiation or heating of fats or model substances produces oxygen containing fragments (Drawert, 1973), to some extent like those of enzymic breakdown, as well as numerous hydrocarbons (Table 2, 3). Hence the reaction mechanisms that produce these products could be at least partly alike, for instance radical reactions.

Our conception of the biological origin of plant aromas is summarized in Figures 3–5 (Tressl et al., 1970).

Known pathways are almost adequate to explain the formation of fatty acids, alcohols, esters and methylketones. Terpenes and terpene-like structures are known to play an extraordinarily large role among aroma substances. They provide classical examples of the fundamental relation of chemical structure and biological effect. Many terpenes display sensory properties and related physiological characteristics

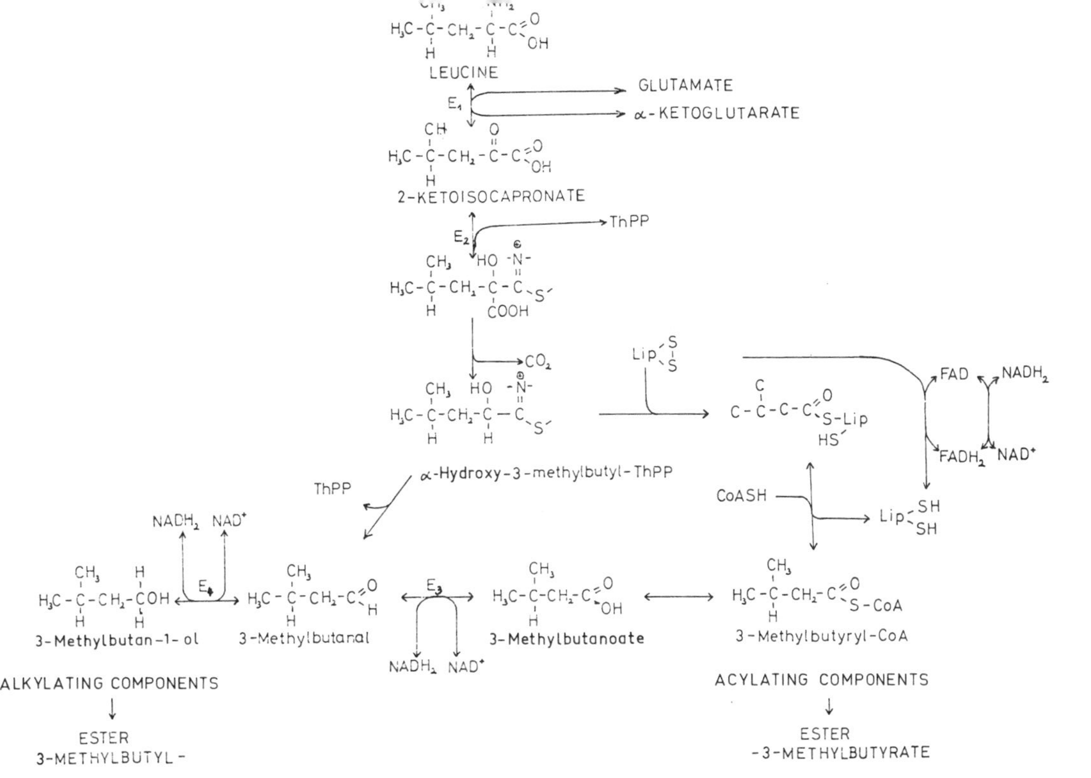

Fig. 3. Conversion of amino acids into aroma components of banana as illustrated by leucine. E_1, L-leucine aminotransferase (EC 2.6.1.6); E_2, pyruvate decarboxylase (EC 4.1.1.1); E_3, aldehyde dehydrogenase (EC 1.2.1.3); E_4, alcohol dehydrogenase (EC 1.1.1.1); ThPP, thiamine pyrophosphate; Lip<S,S>, oxidized lipoic acid; Lip<SH,SH>, reduced lipoic acid; FAD, flavin-adenine dinucleotide; NAD^+, oxidized nicotinamide-adenine dinucleotide; CoA-SH, coenzyme A.

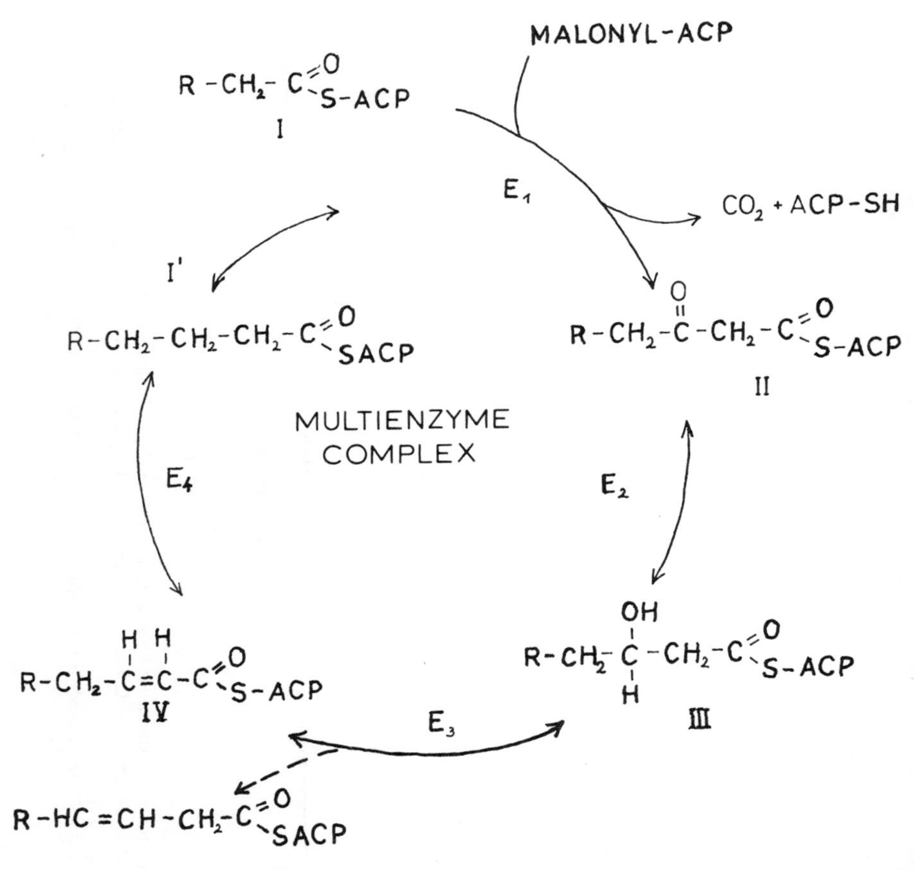

Fig. 4. Fatty acid spiral in plants: splitting off of aroma components. From Compound I, saturated fatty acids; from II + III, methylketones and secondary alcohols; from IV, unsaturated fatty acids, unsaturated aldehydes, unsaturated alcohols, and γ-lactones. ACP, acyl carrier protein; E_1, acyl-ACP: malonyl-ACP ligase; E_2, β-oxoacetyl-ACP:NADH oxidoreductase; E_3, D-β-hydroxyacyl-ACP:NAD(P) oxidoreductase; E_4, α,β-dehydroacyl-ACP:NADH(NADPH) oxidoreductase.

with a certain structural specificity. Hops are known to contain many terpenoids in their aromatic oils, as well as the bitter substances.

Figure 6 shows a selection of monoterpenes and sesquiterpenes found in hop oil. They can be explained as formed by synthesis from prenyl residues or by contribution of prenyl residues. The hop plant seems to be generally inclined towards prenylation reactions. Since Riedel and his colleagues at our institute demonstrated the structure of humulone and lupulone by total synthesis and since bitter substances are of considerable importance in the food and beverage industry, we examined how hop bitter substances could be formed in the hop plant. Figure 7 lists the most common bitter substances in hop.

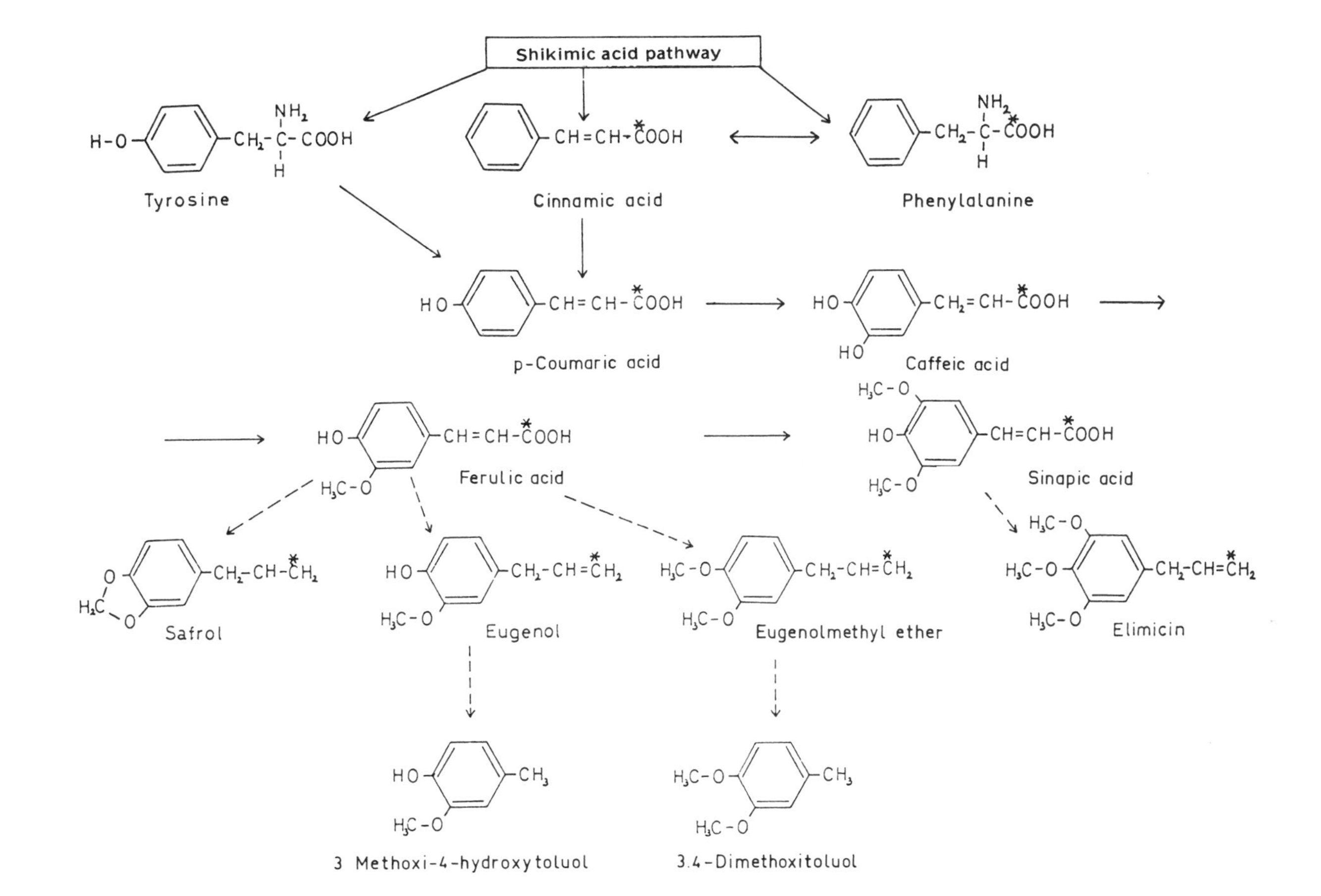

Fig. 5. Synthesis of phenol ethers in banana.

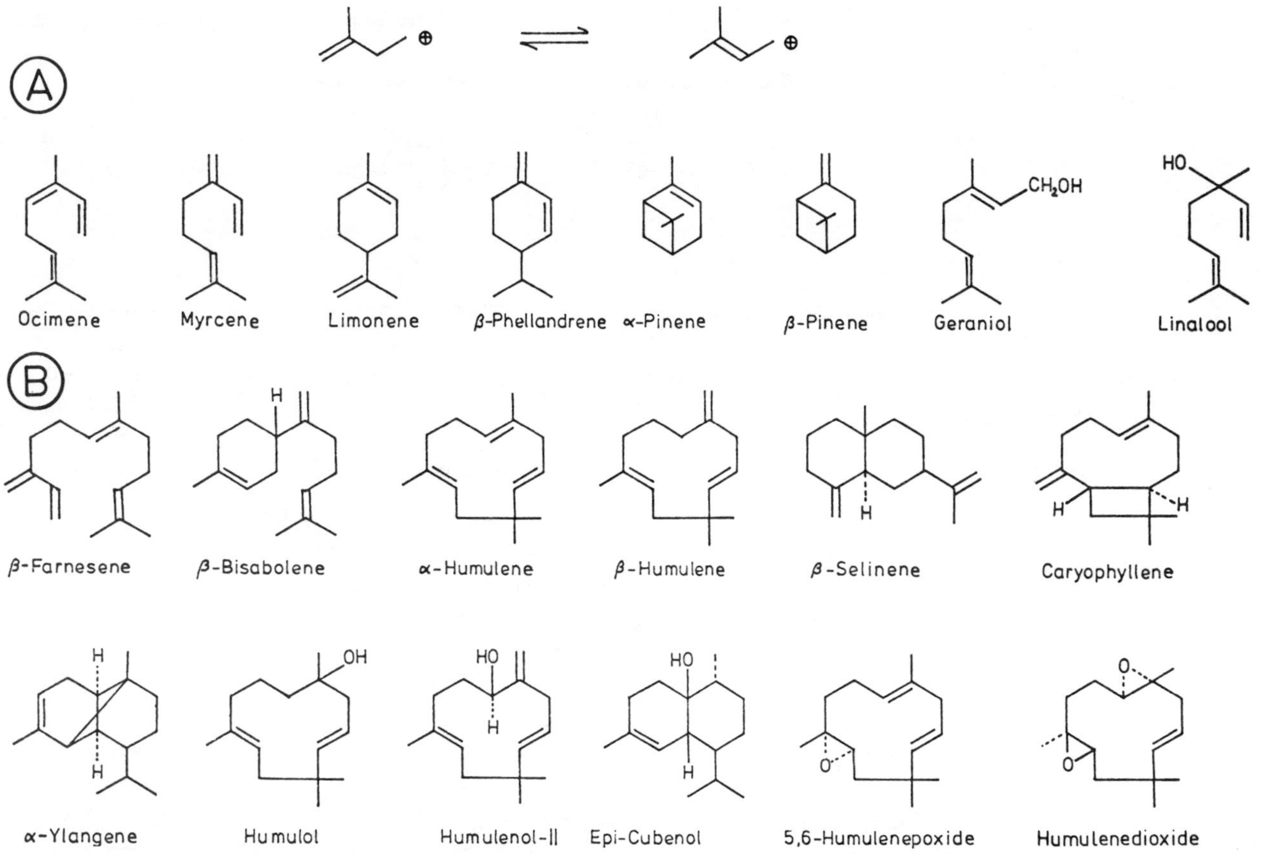

A

Ocimene
Myrcene
Limonene
β-Phellandrene
α-Pinene
β-Pinene
Geraniol
Linalool

B

β-Farnesene
β-Bisabolene
α-Humulene
β-Humulene
β-Selinene
Caryophyllene
α-Ylangene
Humulol
Humulenol-II
Epi-Cubenol
5,6-Humulenepoxide
Humulenedioxide

Fig. 6. Terpenes detected in hop oil. A: monoterpene precursors, B: sesquiterpenes.

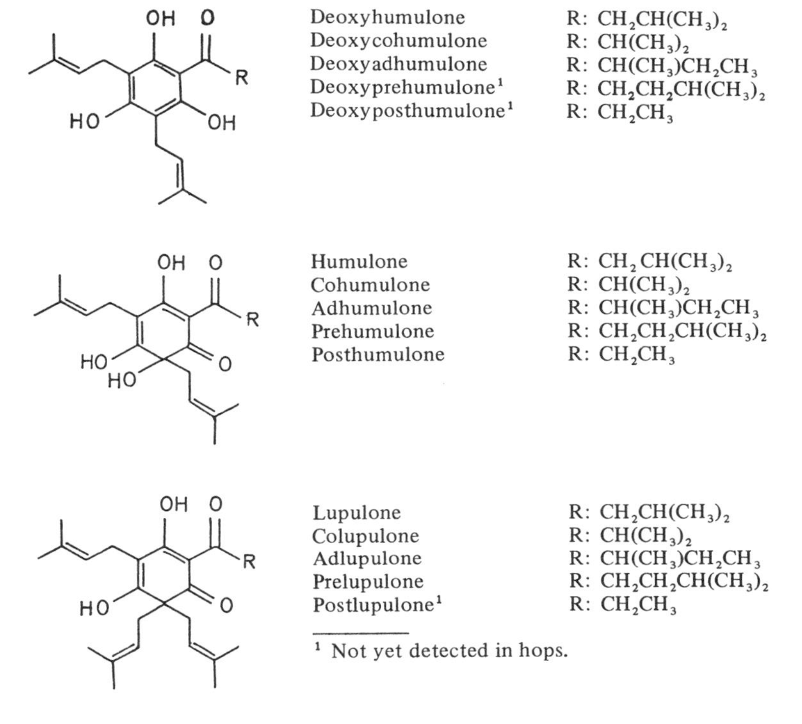

Fig. 7. Scheme of the main bitter substances of hops.

If we look upon the structural formulae of hop bitter substances (Fig. 7), we can distinguish 3 structural elements.

1. The nucleus is a 6–C ring, either aromatic or of cyclohexadiene form, with various hydrogen atoms attached.
2. The terpenoid side-chain is alike in all three groups of bitter substances and can be considered as prenyl residues (3,3-dimethylallyl or 3-methylbut-2-enyl). Deoxyhumulone and humulone include prenyl groups on C–3 and C–5 of the ring, and lupulone contains one more prenyl residue on C–3.
3. The acyl side-chain substances distinguished within each group of bitter compounds, e.g. the lupulone group, includes isobutyryl, isovaleryl and 2-methylbutyryl derivatives called colupulone, lupulone and adlupulone respectively.

We can now discuss whether the ring is built first and then substituted or whether the acyl chain is synthesized before folding and ring formation. From the evidence, we suspected the acyl chain to be formed first, especially since Mori found that about 13 days after flowering leucine decreases. At this same time, bitter compounds are being formed in the lupulin glands. Hence leucine can, after transamination, serve as precursor for the acyl side-chain. We suspect that isovalerate is the precursor for the acyl side-chain of lupulone, humulone and deoxyhumul-

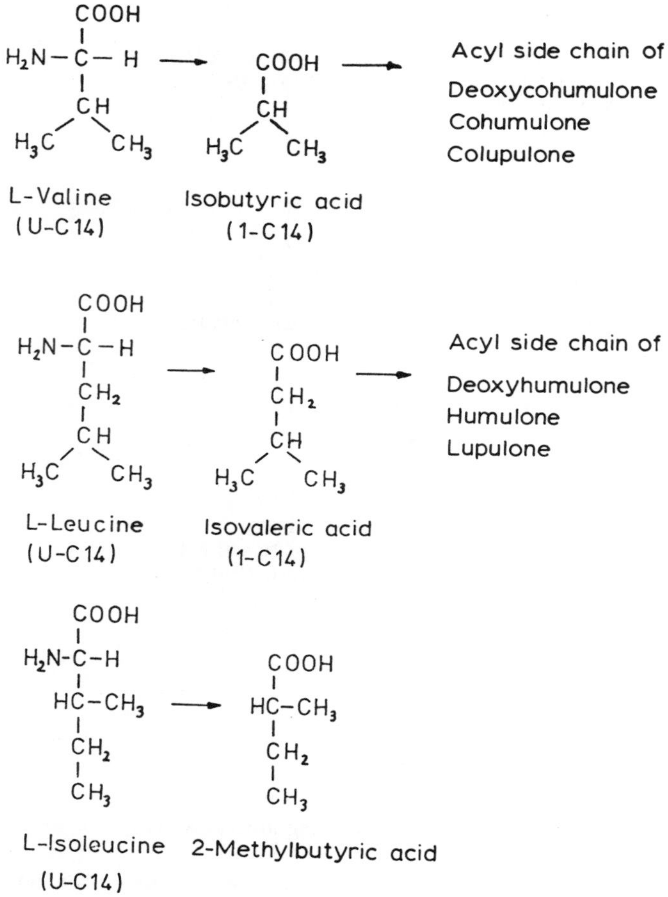

$$H_2N{-}CH(COOH){-}CH(CH_3)_2 \longrightarrow (CH_3)_2CH{-}COOH \longrightarrow$$

L-Valine (U-C14) → Isobutyric acid (1-C14) → Acyl side chain of Deoxycohumulone, Cohumulone, Colupulone

$$H_2N{-}CH(COOH){-}CH_2{-}CH(CH_3)_2 \longrightarrow (CH_3)_2CH{-}CH_2{-}COOH \longrightarrow$$

L-Leucine (U-C14) → Isovaleric acid (1-C14) → Acyl side chain of Deoxyhumulone, Humulone, Lupulone

$$H_2N{-}CH(COOH){-}CH(CH_3){-}CH_2{-}CH_3 \longrightarrow HOOC{-}CH(CH_3){-}CH_2{-}CH_3$$

L-Isoleucine (U-C14) → 2-Methylbutyric acid

Fig. 8. Incorporation of radioactively labelled compounds.

one. Indeed after supplying of unlabelled propionate and isovalerate to whole hop plants, the content of humulone increased.

Then we introduced the labelled compounds listed in Figure 8 into the stem of hopvines with about 12 hops and 3 leaves. The stems were cut under water. After 8 h in the open exposed to sunlight, all the parts were homogenized with methanol and the substances of interest, partly from the residue of evaporation, partly from pentane extraction, were silylated with hexamethyldisilazan in dimethyl-formamide. Only the reaction-radio-gas chromatography in an all-glass system yielded unambiguous results. With [U-^{14}C]L-leucine, we demonstrated the transformation of L-leucine to isovalerate and further transformation from 3-hydroxy-3-methylglutarate to acetoacetate and acetate.

Our results showed that L-leucine was incorporated into deoxyhumulone and lupulone. We assume that the incorporation occurs as isovaleryl side-chain into the bitter substances. Our grounds are as follows. We could detect the transformation of L-leucine to isovalerate, which we consider as the real precursor of the acyl side-chain. Introduction as prenyl side-chain (3-hydroxy-3-methylglutarate – mevalonate – deoxyhumulone or lupulone) seems most unlikely. We found no radioactivity in deoxycohumulone, colupulone or in an other bitter substance. It would have been expected at least in colupulone, which is present in larger amounts than lupulone in the hop variety used. Further there was no evidence that L-leucine was built into the terpenes or phytosterins in the hops. If mevalonate were formed from the L-leucine, activity would have appeared in squalene, since under the experimental conditions we found that mevalonate was preferentially incorporated in squalene. Since this was not so, we can conclude that the acetoacetate and acetate derived from L-leucine can likewise not be precursors of prenyl residues, though they could also be incorporated in the ring of the bitter substances. If this were so, we would expect other bitter compounds to be labelled, at least colupulone, since the C_6 ring is common to all. Since this is not so, the incorporation of L-leucine specifically in deoxyhumulone and lupulone can be easily explained in terms of derivation of the acyl side-chain from isovalerate.

Test with [1-^{14}C]isobutyrate showed incorporation in deoxyhumulone, cohumulone, and colupulone; [1-^{14}C]isovalerate was incorporated in deoxyhumulone, humulone and lupulone. We also found an increased incorporation of the two carboxylic acids, indicating that they are precursors of the acyl side-chain. In the experiments with labelled isobutyrate and isovalerate, we found also in the radiograms substances of relatively high activity which only justs differ in gas-chromatographic behaviour. They proved to differ only in the acyl residue, isobutyryl and isovaleryl. These conclusions led to the recognition of two universal precursors. We called these compounds CoX and X (Fig. 9).

We confirmed the structure of CoX and X by total synthesis. One can imagine that once these compounds are formed simple reactions of CoX and X yield the

1-Isobutyryl-3-(3,3-dimethyl-allyl)-phloroglucin

CoX

1-Isovaleryl-3-(3,3-dimethyl-allyl)-phloroglucin

X

Fig. 9. Universal precursor of hop bitter substances.

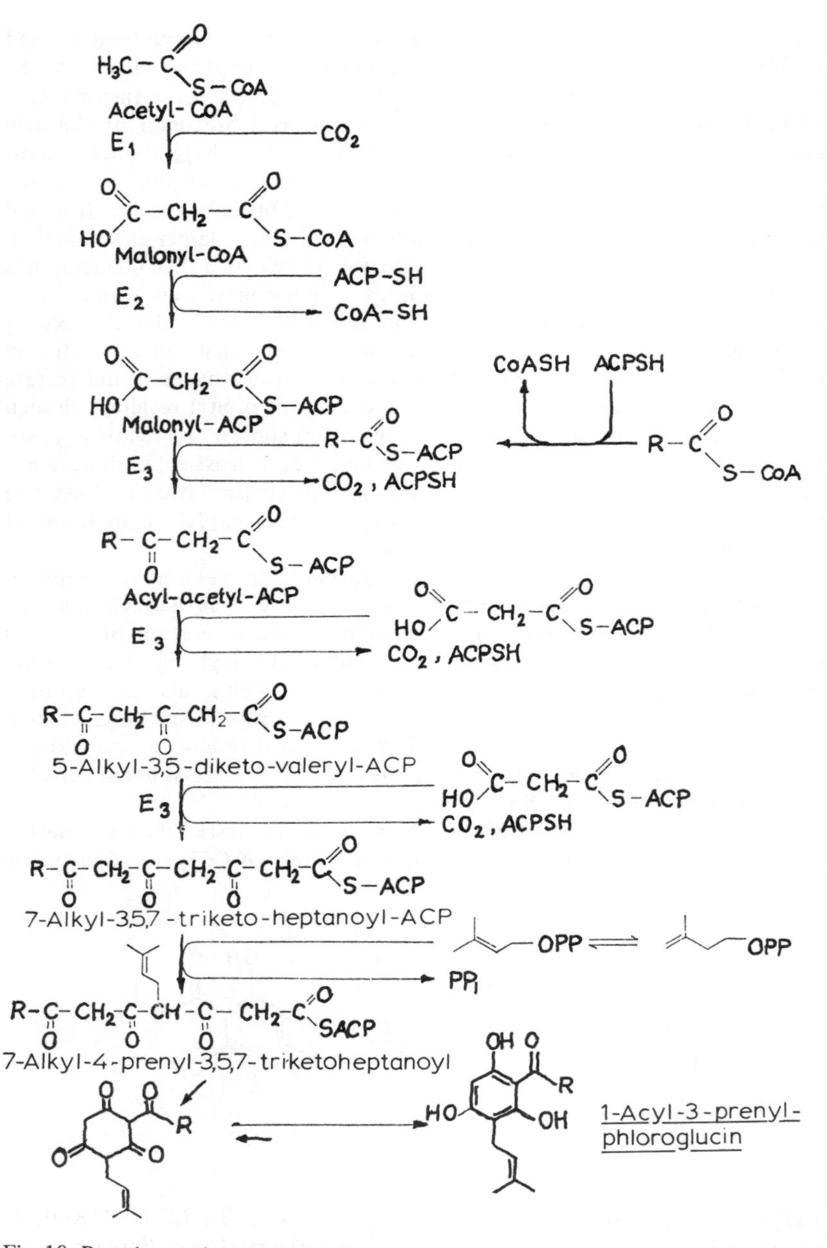

Fig. 10. Reaction pathways for synthesis of bitter substances in hops. E_1, acetyl-CoA carboxylase; E_2, malonyl transacylase; E_3, acyl-ACP:malonyl-ACP ligase; CoA-SH, Coenzyme A; ACP, acyl carrier protein.

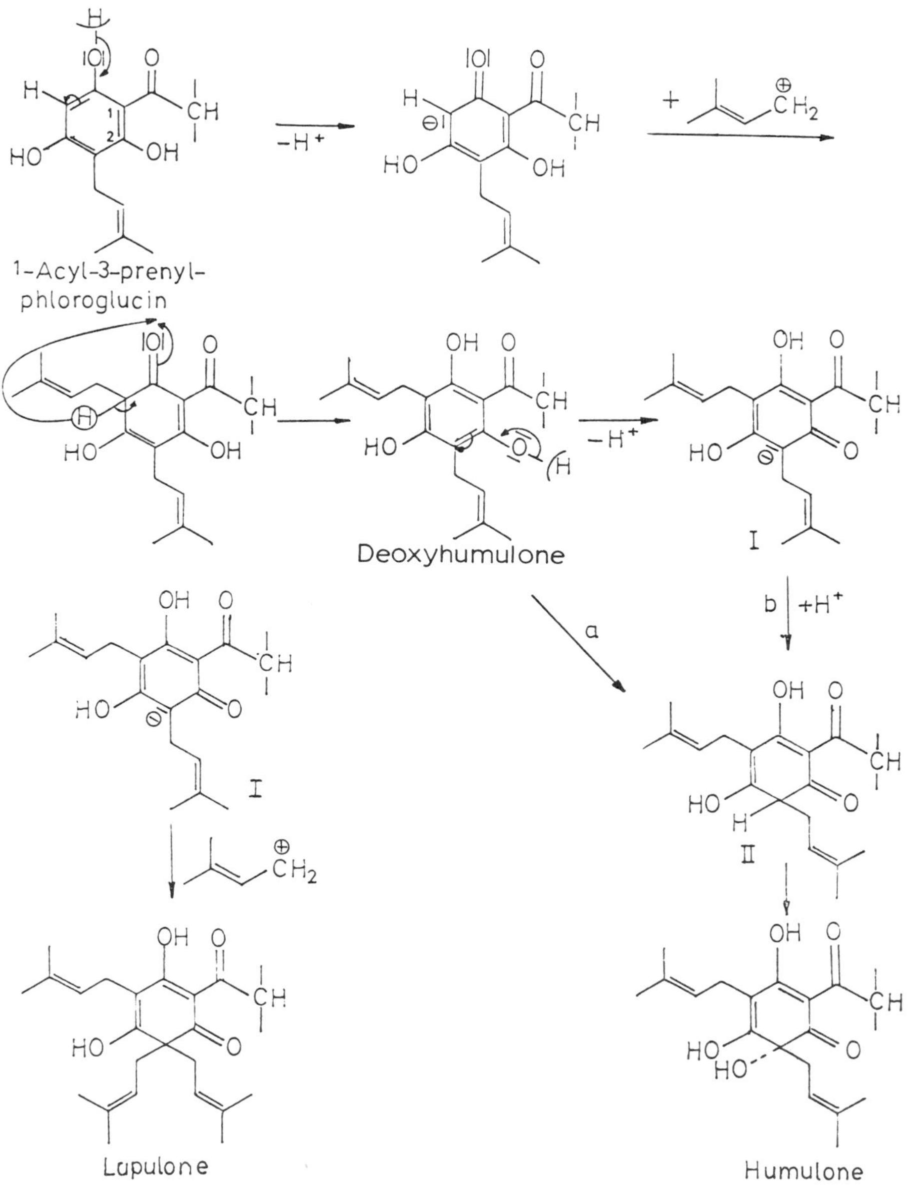

Fig. 11. Reaction pathways for synthesis of deoxyhumulone, humulone and lupulone from the central intermediary.

Co-components and the skeleton of the bitter substances. Single or double prenylation of the compound CoX would produce deoxycohumulone and colupulone, respectively. Monoprenylation and hydroxylation would yield cohumulone.

After proving the formation of isobutyrylacetate from [^{14}C]acetate and [^{14}C]isobutyrate, we proposed the synthetic pathway in Figure 10 for bitter compounds in hops. The pathway begins with condensation of the carboxylate and malonate of the acyl side-chain. With carboxylase E_1, acetylcoenzyme A is transferred to malonylcoenzyme A, from which malonyl-ACP is formed with the action of transacylase E_2 and ACP.SH, where ACP is Acyl Carrier Protein, here in the sulphydryl form. With the enzyme E_3, a ligase, the carboxyl residue (e.g. isobutyryl or isovaleryl residue) is attached, carbon dioxide being split off, forming a further energy-rich intermediate, acylacetyl.ACP, which can again react with malonyl.ACP to form 4-acyl-3-ketobutyryl.ACP, in other words 5-alkyl-3,5-diketovaleryl.ACP. One could imagine that prenylation directly ensues at this stage, but steric evidence conflicts with this. More likely is a further extension of the chain to form 7-alkyl -3,5,7-triketoheptanoyl.ACP where the alkyl may be either isopropyl or isobutyl. Figure 10 shows that this compound could be now prenylated, particularly since further condensation of malonyl.ACP does not occur. Ring formation can be envisaged as a splitting off of the $ACPS^-$ anion and a proton from C–6. The assumed intermediate cyclohexa-2,4,6-trione structure tautomerizes to an aromatic structure, 1-isobutyryl and isovaleryl-3-prenylphloroglucin, which we identified as the central precursors CoX and X (Fig. 11).

The 1-acyl-3-prenylphloroglucin is then stepwise further prenylated. For steric reasons, the second prenyl group is probably easier incorporated at C–5 than at C–3, since C–3 already has a 3,3-dimethylallyl residue. In a way, one can view the deoxyhumulone formed as another universal precursor for lupulone and humulone. The third prenylation is a likely at C–3 as C–5. We can then envisage the formation of lupulone and humulone either by tautomerization from deoxyhumulone (Path a) to Compound II or via the anion to Compound II (Path b). Compound II is transformed by *o*-hydroxylation to humulone, such *o*-hydroxylation being a commonly observed reaction in higher plants (Drawert & Beier, 1973, 1974a, b, c).

Relationships are more complex when different pathways share in the creation of an aroma. This is true, for instance, of alcoholic beverages like cider, wine, or

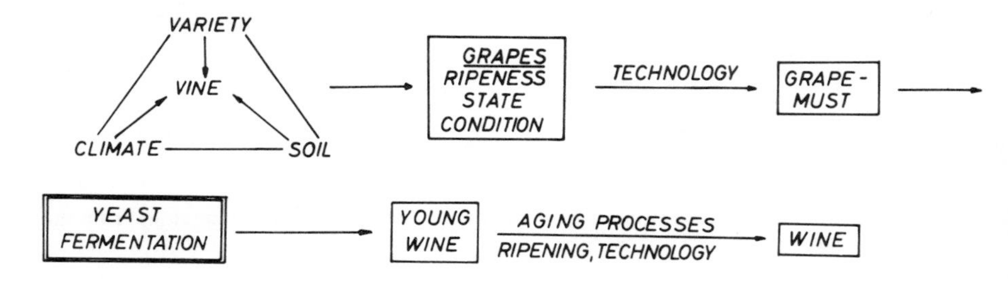

Fig. 12. Biological-technological sequence of wine.

Table 4. Dependence of the concentration of free amino acids (mg/l) in two vine varieties on amount of nitrogen fertilizer (kg/ha). All treatments were dressed with potassium and phosphorus each at 2 kg/ha.

Fertilizer		Total amino acids	His	Lys	Arg	NH_3	Asp	Thr	Ser	Glu	Pro	Gly	Ala	Val	Met	Ileu	Leu	Phe
N	nitrate (N_2O_5)																	
Aris[1]																		
0.5	3.5	960	+	+	480	29	39	26	128	222	+	+	63	19	+	12	20	22
2.0		1598	20	+	801	39	22	38	204	203	+	+	107	27	15	25	35	62
4.0		2651	15	+	853	82	57	44	470	216	480	7.5	252	27	21	31	50	66
6.0	29.0	2972	36	+	828	87	70	88	495	268	590	4.5	318	46		31	45	67
Riesling																		
0.5	1.5	800	16	+	174	10	49	30	65	132	123	+	86	22	21	26	33	31
2.0	2.0	1064	23	+	314	20	36	49	78		136	7.5	136	40	30	31	42	48
4.0	4.4	2445	48	13	810	53.0	63	123	305	370	179	22.5	454	45	24	23	32	49

1. New vine variety.

beer. Fermented drinks often contain some hundreds of aroma components. Their formation can perhaps best be understood, if we take the aroma as the end-product of a 'biological' or 'biotechnological sequence' (Drawert & Rapp, 1966; 1967). Figure 12 shows the biological sequence for wine.

This sequence begins in the plant, being determined by species and variety, and its metabolism is influenced by conditions such as climate, soil and even fertilizers (Drawert, 1974). Accordingly the components of grapes vary during development.

Table 4 shows the effect of nitrogen dressing on the content of free amino acids in the grapes. Alongside ripening, grape processing, mashing and their importance for the development of secondary aroma components, fermentation is particularly affected by the amino acids. Nitrogen compounds enter the yeast's metabolism, so

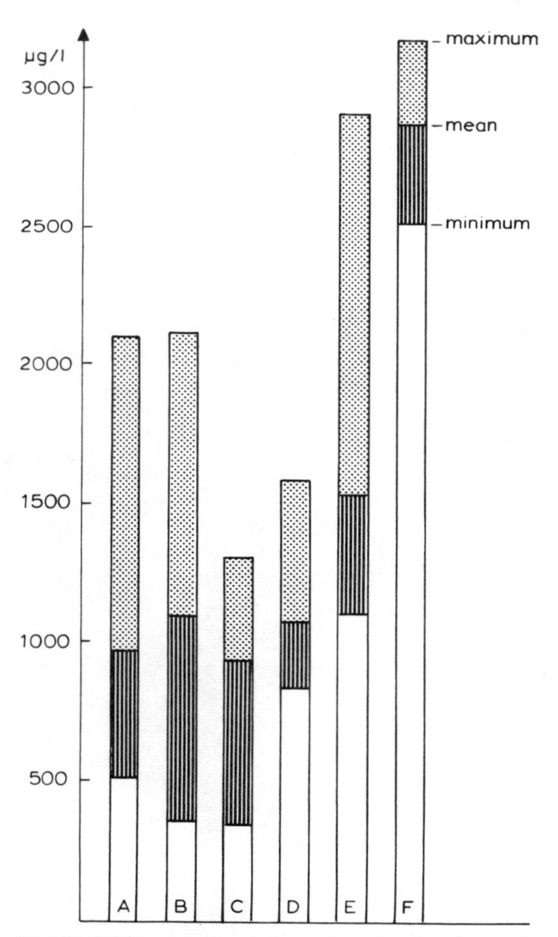

Fig. 13. Mass concentration of 3-methylthiopropan-1-ol in different types of beer. A: clear Lager; B: clear export beers; C: Pils beers; D: dietary beers; E: old-beers; F: wheaten beers.

to say, with 'biochemical valence' in that they determine which aroma substances are formed during fermentation.

We extended our earlier model trials on fermentation, to examine the formation of volatile aroma components, with studies of methionine metabolism in yeast. We demonstrated that methionine is essentially metabolized to methionol (3-methylthiopropan-1-ol), a product of the Ehrlich reaction, whereby traces of other sulphur-containing components are formed such as 3-methylthiopropionate and its ethylester and 3-methylthiopropyl acetate.

We estimated these sulphur compounds quantitatively in fermented beverages and discovered that the concentration of methionol in wine and beer was of the order mg·litre^{-1} (Fig. 13, 14), the same range as sulphur compounds of known

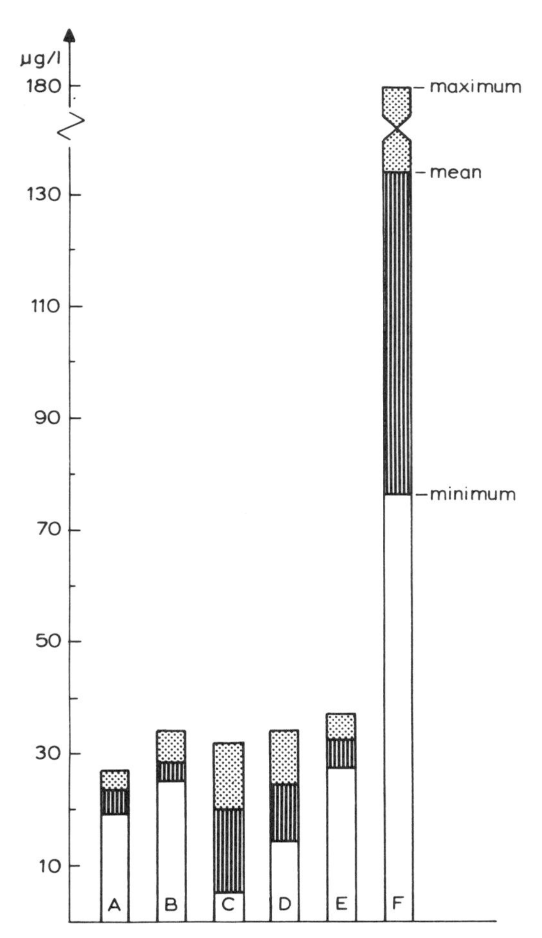

Fig. 14. Mass concentration of 3-methylthiopropyl acetate in different types of beer. A: clear Lager; B: clear export beers; C: Pils beers; D: dietary beers; E: stouts; F: wheaten beers.

importance for taste and smell (Schreier et al., 1974, 1975).

Exact knowledge of the different members of a 'biological sequence' and the linkage of biochemical reactions provide a basis for biotechnology, in the sense that variation of the substrate and regulation of the conditions can radically alter the end-product.

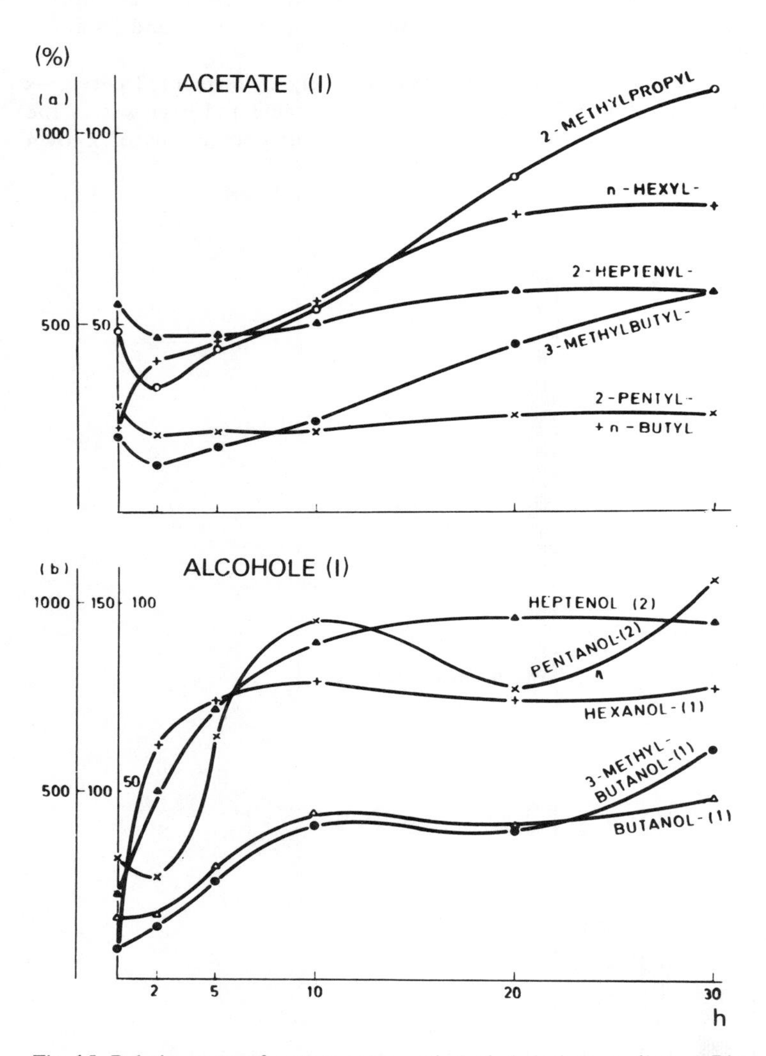

Fig. 15. Relative mass of acetate esters and alcohols in banana discs at Ripeness Stage I. Ordinate, percentage of internal standard. The upper graph values on left apply to *n*-butyl + 2-pentyl, and 3-methylbutyl acetate; on right to 2-methylpropyl, *n*-hexyl, and 2-heptenyl acetate. On lower graph, values on left apply to 3-methylbutan-1-ol and 1-hexanol; in the centre to 2-pentanol; on the right, to 1-butanol, and hept-*cis*-4-en-2-ol.

This presentation also indicates that addition of certain precursors during a biological production process may influence aroma more than addition of the aroma substances themselves. This view and approach also allows us to exploit biochemical and microbiological reactions for the stereoselective synthesis of components with intense aroma properties. As illustrated by the 'biological sequence' of

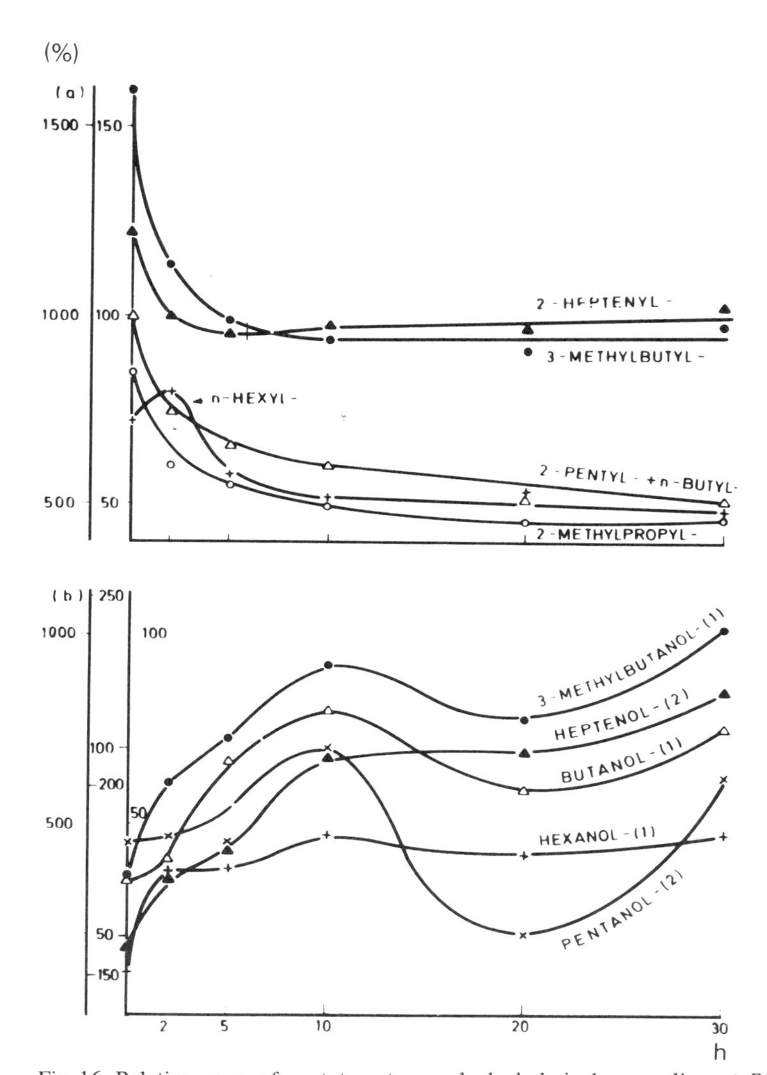

Fig. 16. Relative mass of acetate esters and alcohols in banana discs at Ripeness Stage III. Ordinate, percentage of internal standard. On upper graph, values on left apply to 2-methylpropyl, *n*-butyl + 2-pentyl, and 3-methylbutyl acetate; on right to *n*-hexyl, and 2-heptenyl acetate. On lower graph, values on left apply to 3-methylbutan-1-ol, and 1-hexanol; centre left to 2-pentanol; centre right to hept-*cis*-4-en-2-ol; on right to 1-butanol.

wine, the interaction of a series of biosynthetic pathways yields numerous components of low aroma value and only a few of them contribute notably to the total aroma. In different varieties of grape, people have tried to define characteristic aromas, for instance by trying to relate the main byproducts of fermentation such as fusel oil alcohols or esters with the aroma characteristic of a variety. This search has hardly had any success, since differences between grape varieties are essentially quantitative and since the differences are almost certainly due to minor components present only in traces. But the differences, for instance, between Traminer Riesling and Burgundian wines are clearly detectable by the senses.

Our trials over some years show that the differences between grape varieties are largely attributable to terpenes. Examples are linalool, nerol, geraniol, α-terpineol and 4-terpinenol as well as the aroma components 3,7-dimethyl-octa-1,5,7-trienol (Ho-trienol), the linalool oxides in its furanoid and pyranoid forms, and 2-vinyl-2-methyltetrahydrofuran-5-one, nerol oxide and the isomeric rose oxides (Schreier & Drawert, 1974).

In closing, let me indicate some ways of biotechnologically producing aromas. If banana slices (Cavendish) 5–6 mm thick are incubated in 0.4–M sucrose solution,

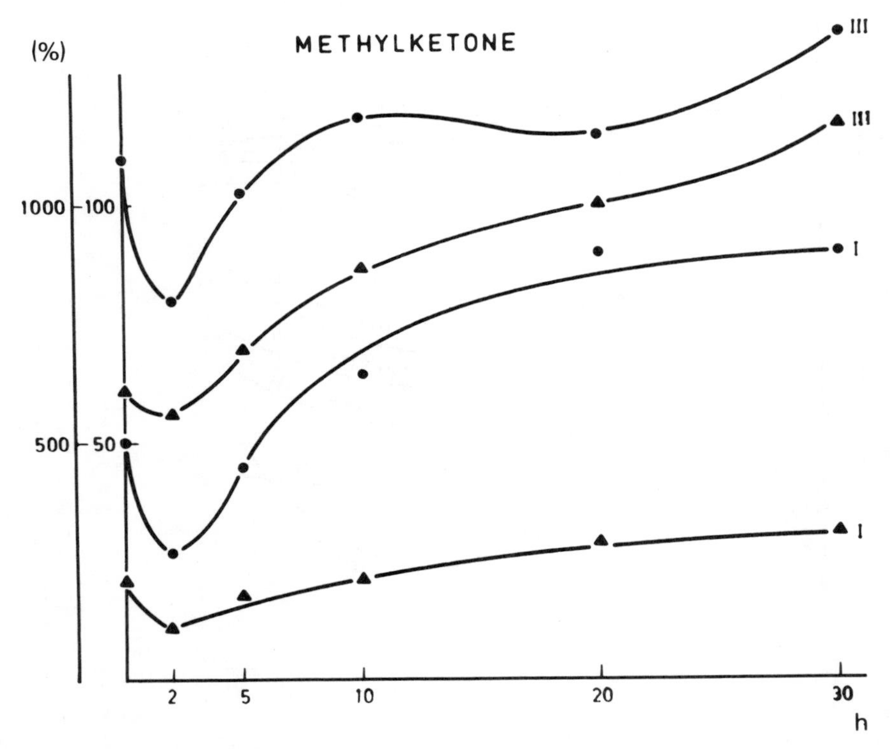

Fig. 17. Relative mass of methylketones in banana discs at Ripeness Stages I and III. Ordinate, percentage relative to internal standard. Values on left apply to 2-pentanone (● — ●) and those on right to 2-heptanone (▲ — ▲).

pH 5.2, at different climacteric stages of ripeness, initially acetate, butyrate and alcohols decrease as a result of slicing and incubation. After 2 h, significant amounts of aroma substances are formed. Figures 15 and 16 show the relation of acetate and alcohols at the climacteric stage of ripeness (I) and after the climacterium, when amounts of aroma substances are decreasing (III).

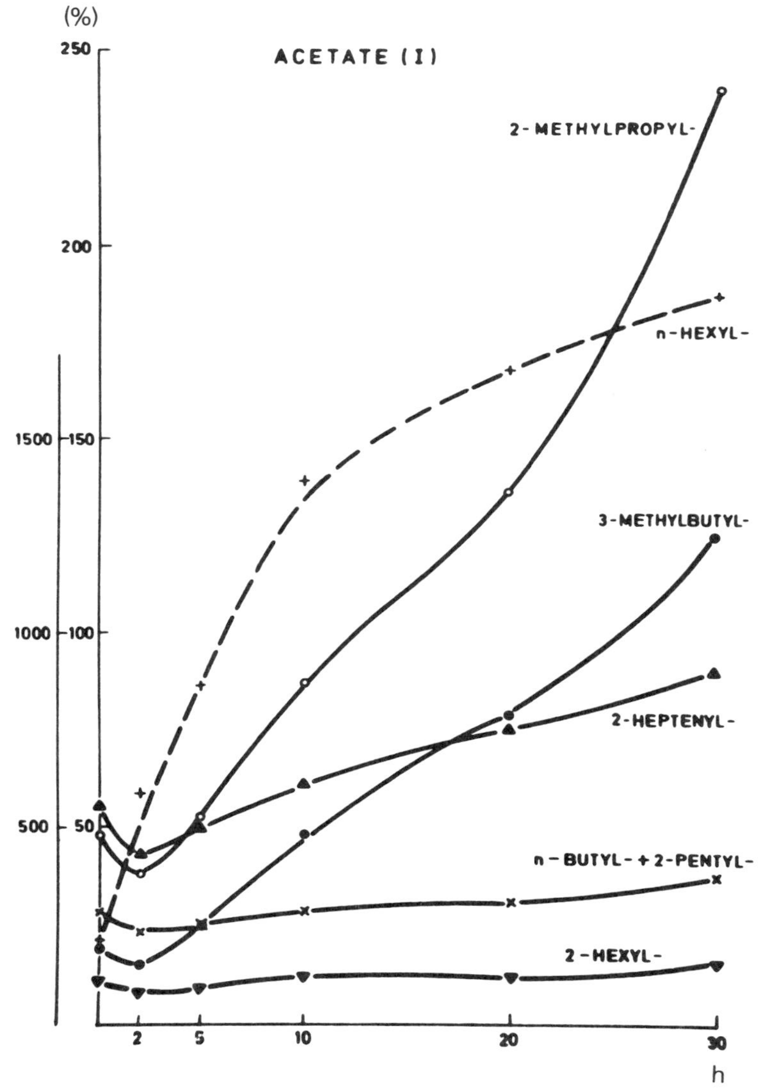

Fig. 18. Incorporation of exogenous acetate at Ripeness Stage I. Ordinate as in Fig. 15. Values on left apply to *n*-butyl + 2-pentyl (x — x), and 3-methylbutyl (● — ●); on right to 2-methylpropyl (○ — ○), 2-hexyl (▼ — ▼), *n*-hexyl (+ — +) and 2-heptenyl acetate (▲ — ▲).

Methylketones such as 2-pentanone and 2-heptanone were synthesized at different rates according to stage of ripeness (Fig. 17) (Drawert & Künanz, 1975).

Exogenous precursors were also partially converted to aroma substances. As Figures 18, 19 and 20 show, tissue slices are more suitable than homogenates for biosynthesis of aroma substances. In the homogenates of banana, breakdown reac-

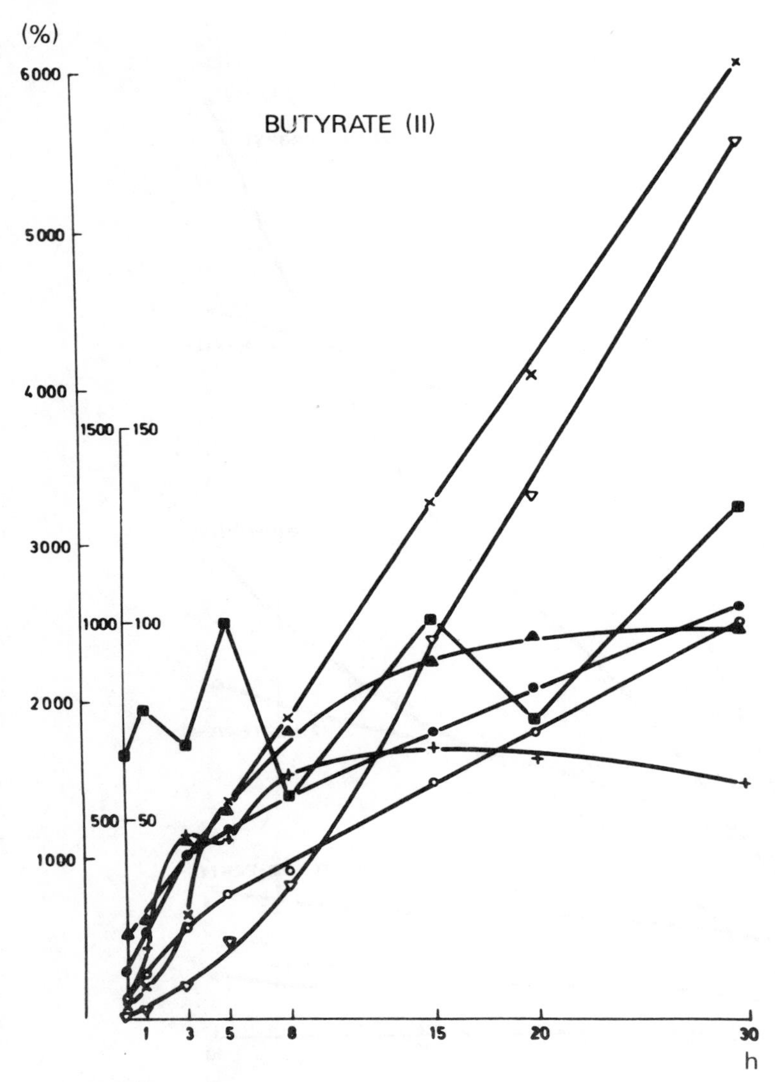

Fig. 19. Incorporation of exogenous butyrate. Ordinate as in Fig. 15. Values on left apply to ethyl (▽ — ▽), *n*-butyl + 2-pentyl (x — x), and 3-methylbutyl butyrate (• — •); in centre to methyl (▪ — ▪), 2-methylpropyl (○ — ○), and *n*-hexyl (+ — +); on right to 2-heptenyl butyrate (▲ — ▲).

tions predominated over synthesis.

As is known, many microbes form aroma substances in suitable substrates and at suitable stages of culture. We have continued trails with various microbes in different media. One of the organisms we used was *Ceratocystis moniliformis,* whose ability to form aroma substances was demonstrated by Jim Palmer and his colleagues at the 1972 symposium of the American Chemical Society in New York.

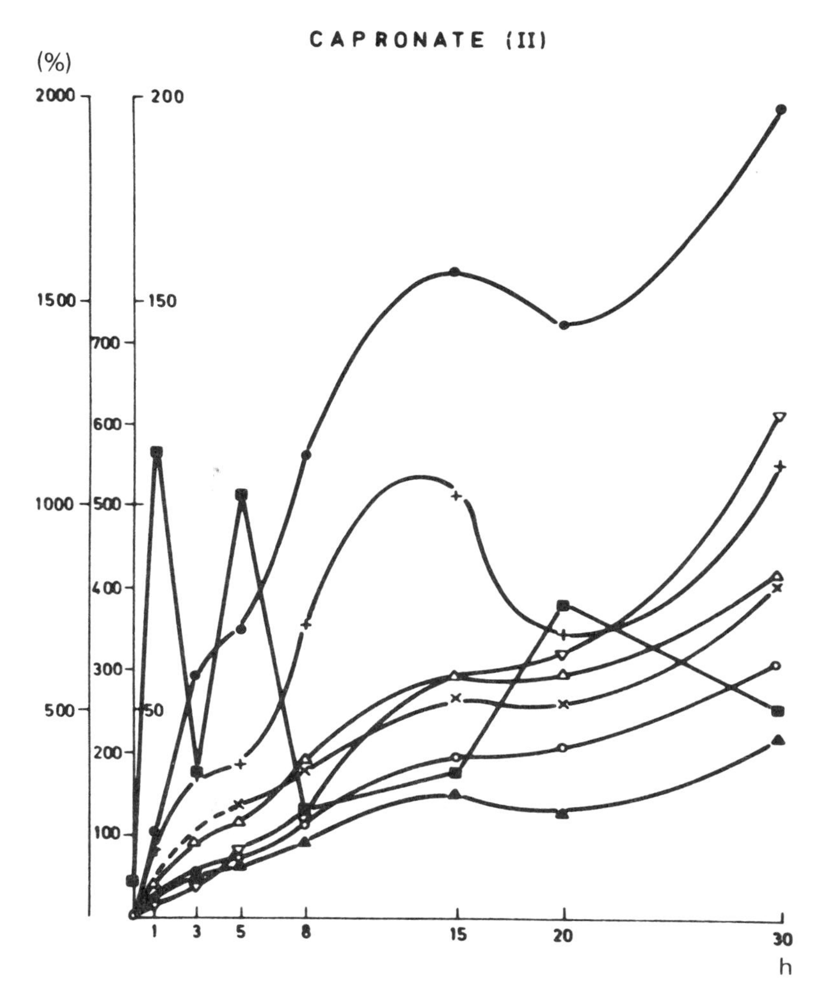

Fig. 20. Incorporation of exogenous hexanoate. Ordinate as in Fig. 15. Values on left apply to ethyl (▽ — ▽), 2-methylpropyl (○ — ○), *n*-butyl (△ — △), 3-methylbutyl (● — ●), and *n*-hexyl hexanoate (+ — +); in the centre, to methyl (■ — ■); on right to 2-pentyl (x — x), and 2-heptenyl hexanoate (▲ — ▲).

Table 5 summarizes the aroma substances so far isolated and identified from the culture solution. As you will see, this microbe can synthesize a whole series of structures, including terpenes (Drawert & Barton, in preparation).

Table 5. Identified aroma substances isolated from a liquid culture of *Ceratocystis moniliformis.*

2-Phenethyl acetate	α-Terpineol
(Isobutanol)	Geraniol
3-Methylbutan-1-ol	2-Methylhept-2-en-6-one
2-Phenylethanol	2-Nonanone
Isopentyl acetate	Propyl acetate
a C_{10}-Methyl ester	Isobutyl acetate
a C_{11}-Methyl ester	Geranyl acetate
δ-Nonalactone	Neryl acetate
δ-Dodecalactone	a C_7-Methyl ester
Hex-*trans*-3-en-1-ol	a C_8-Methyl ester
1-Heptanol	γ-Nonalactone
Linalool	2-Methylhept-2-en-6-ol
	Nerol

References

Amoore, J. E., 1962. Proc. Sci. Sect. Toilet Goods Assoc. Suppl. 37, 1: 13.
Amoore, J. E., 1963. Nature (London) 198: 271.
Amoore, J. E., 1967. Nature (London) 214: 1095.
Belitz, H.-D., 1973. Struktur und Geschmack-Bitterpeptide. In: K. Heyns (ed.), Aktuelle Probleme der Lebensmitteltechnologie. Vorträge wissenschaftlicher Arbeitstagung, Institut für Forschung und Entwicklung der Maizena Gesellschaft m.b.H. Z. Ernaehrungswiss. Suppl. 16, p. 150–158.
Drawert, F., 1973. Proc. int. Coll. Die Identifizierung von bestrahlten Nahrungsmitteln. Bundesforschungsanstalt für Ernährung, Karlsruhe, October 1973. Euratom EUR 5126 d, e, f, i, n, p. 109.
Drawert, F., 1974. In: A. D. Webb (ed.), Chemistry of Winemaking, Adv. Chem. Ser. 137. American Chemical Society Washington DC.
Drawert, F. & H. Barton. In preparation.
Drawert, F. & J. Beier, 1973. Brauwissenschaft 26: 357.
Drawert, F. & J. Beier, 1974a. Chromatographia 7: 273.
Drawert, F. & J. Beier, 1974b. Phytochemistry 13: 2149.
Drawert, F. & J. Beier, 1974c. Phytochemistry 13: 2749.
Drawert, F. & H. J. Künanz, 1975. Chem. Mikrobiol. Technol. Lebensm. 3: 185.
Drawert, F. & A. Rapp, 1966. Vitis 5: 351.
Drawert, F. & R. Tressl, 1970. Ernährungsumschau 10: 392.
Drawert, F., A. Rapp & W. Ullemeyer, 1967. Vitis 6: 177.
Drawert, F., W. Heimann, R. Emberger & R. Tressl, 1965. Naturwissenschaften 52: 304.
Drawert, F., W. Heimann, R. Emberger & R. Tressl, 1966. Liebigs Ann. Chem. 694: 200.
Drawert, F., W. Heimann, R. Emberger & R. Tressl, 1969. Z. Naturforsch. 20b: 497.
Hail, G., 1963. Riechstoffe. In: Ullmans Enzyklopädie der technischen Chemie, 3rd edn, vol. 14. Urban & Schwarzenberg, München-Berlin, p. 690–776.
Merkel, D., 1972. Riechstoffe, WTB (Wissenschaftliche Tagebücher). Akademie Verlag, Berlin; Pergamon Press, Oxford; Vieweg & Sohn, Braunschweig.
Moncrieff, R. W., 1951. The Chemical Senses. L. Hill, Ltd, London.
Ohloff, G. & A. F. Thomas, 1971. Gustation and Olfaction. Proc. Int. Symp. Firmenich 75th anniversary. Academic Press, London, New York.

Ruys, A. H., 1962. L'âme des fruits. Berichte 4. Symposium der Internationalen Fruchtsaft-union (Fruchtaromen). Juris Verlag, Zürich.
Schreier, P. & F. Drawert, 1974. Chem. Mikrobiol. Technol. Lebensm. 3: 154.
Schreier, P., F. Drawert & A. Junker, 1974. Z. LebensmUnters. Forsch. 154: 279.
Schreier, P., F. Drawert & A. Junker, 1975. Brauwissenschaft 28: 73.
Shallenberger, R. S., 1971. Molecular Structure and Taste. In: Ohloff & Thomas, 1971, p. 126.
Tressl, R. & F. Drawert, 1973. J. agric. Food Chem. 21: 560.
Tressl, R., F. Drawert, W. Heimann & R. Emberger, 1970. Z. LebensmUnters. Forsch. 144: 4.

Proc. int. Symp. Aroma Research, Zeist, 1975. Pudoc, Wageningen.

Biogenesis of volatiles in fruit and vegetables

R. Tressl, M. Holzer and M. Apetz

Institut für Chemisch-Technische Analyse der Technischen Universität Berlin, 1 Berlin 65, Seestrasse 13, Bundesrepublik Deutschland

Abstract

A few hundred aroma compounds have been identified in fruits and vegetables by combined gas chromatography and mass spectrometry.

Fruit volatiles are esters, alcohols, acids, lactones, carbonyls, terpene and sesquiterpene compounds. The volatiles are produced during postclimacteric ripening by intact cells. Fruit volatiles are derived from amino acid (Leu, Val, Ileu, Met, Phe), fatty acid and terpene metabolism. Some of the biogenetic pathways were proved by labelling experiments with fruit slices and ^{14}C-labelled precursors.

Vegetables produce smaller amounts of volatiles. Intact cells of vegetables contain non-volatile precursors (alkylcysteine sulphoxides, thioglucosides, linoleic and linolenic acids etc.) which are split into volatile components by enzymic reactions during homogenization. Thio compounds and carbonyls are predominant volatiles in vegetables like garlic, onion, radish, cucumbers and tomatoes. Reaction pathways are discussed.

Introduction

The object of most flavour research has been to identify the volatiles responsible for the characteristic flavour and aroma of foodstuffs. The increasing 'Lists of Volatiles in Food', edited by van Straten & de Vrijer (1973) have shown that adequate methods for separation and identification of traces of these volatile compounds are now available, and are more and more used. A rewarding area of future research will be the assessment of the contribution of identified components to any flavour. The mode of biochemical reactions between functional groups of flavour-bearing compounds and the human receptor cells is still unknown in detail (Boekh, 1974). Capillary gas chromatography and mass spectrometry together with sensory evaluation has resulted in the identification of character-impact substances from many fruit and vegetables. As yet little is known about the pathways and control mechanisms in the synthesis and accumulation of substances. Such knowledge would allow us to enhance the production of desirable flavours and reduce or eliminate the production of off-flavours. Let us try to piece together the fragmentary knowledge, and, sometimes speculatively, update our knowledge on the biogenesis of volatiles in fruit and vegetables.

Biogenesis of volatiles in fruit

The typical flavour compounds of fruit such as apples, pears, bananas, and peaches are not produced during growth, nor are they present at harvest. Flavour in these fruit is produced during a short ripening period, related to the climacteric rise

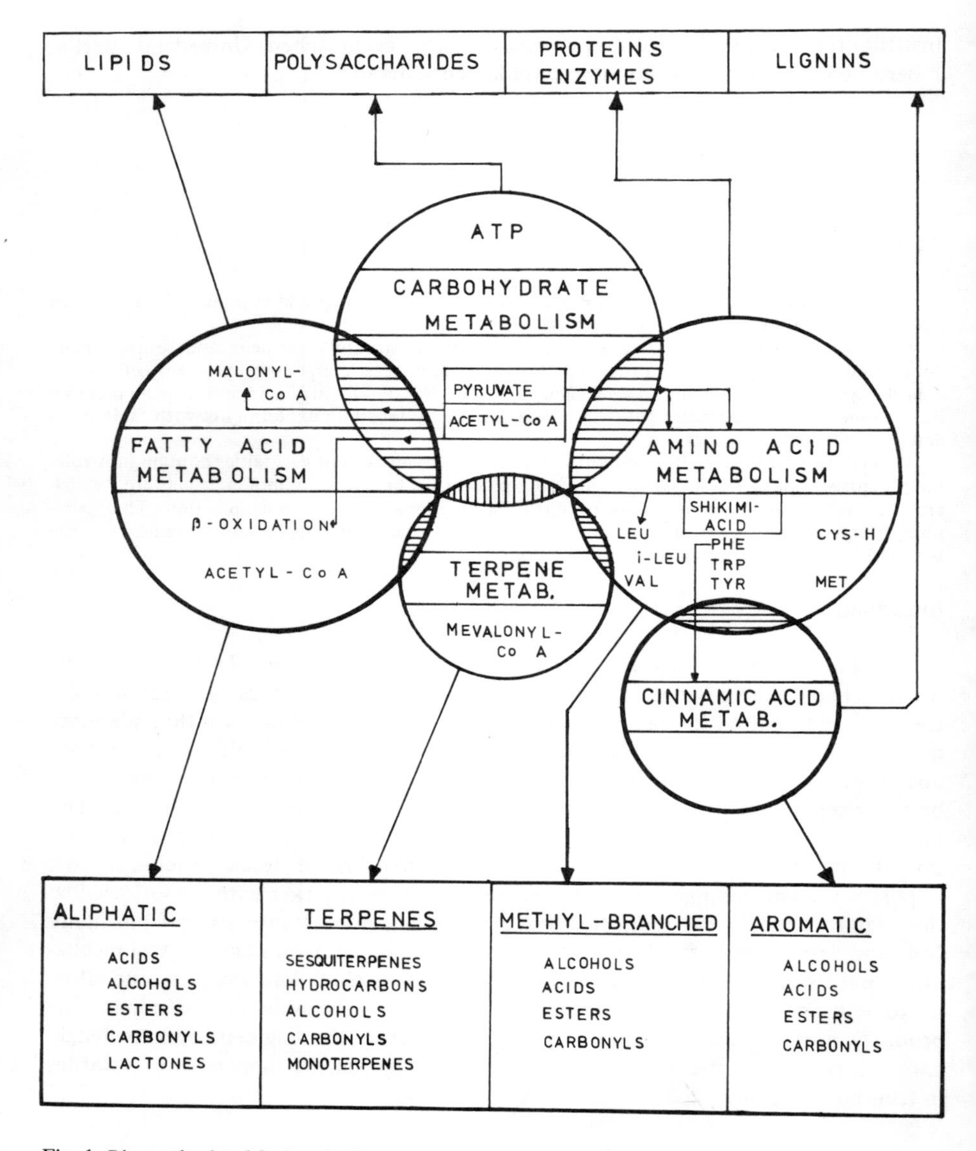

Fig. 1. Biosynthesis of fruit volatiles.

in respiration (Heinz et al., 1965; Romani & Ku, 1966; Drawert et al., 1972; Tressl & Jennings, 1972). Many of these substances are produced by biosynthesis. The growing unripe fruit synthezises macromolecules such as proteins, polysaccharides, lipids and flavonoids by breakdown of carbohydrates, synthezised by photosynthesis in the leaves. The preclimacteric fruit produce small amounts of ethylene which is recognized as a 'ripening hormone', and is obviously derived from methionine (Liebermann et al., 1966).

Ethylene induces biochemical, physical, and chemical changes in colour, texture, permeability of membranes and an increase in amounts of some proteins and enzymes (Biale, 1964; Hansen, 1966). The metabolism of the fruit changes. Catabolism predominates, decreasing the amounts of polysaccharides, oligosaccharides, and fruit acids. One of these changes is the production of volatiles. Figure 1 summarizes possible pathways in the production of aroma substances in fruit. Some of these proposed pathways were investigated by radioactive labelling experiments with fruit slices at different stages of ripening.

According to Heinz et al. (1965), Romani & Ku (1966), Drawert et al. (1972), Tressl & Jennings (1972) and Jennings & Tressl (1974), the production of volatiles in ripening pears, apples and bananas is initiated by the climacteric rise in respiration, reaching maximum values in the postclimacteric ripening phase. This is shown for some pear volatiles (Fig. 2). Bartlett pears that had been subjected to normal commercial processes (harvested green-ripe and stored for 60 days at 4 °C) were ripened in a glass chamber at 25 °C in continuous flow of filtered air. The entrained volatiles were trapped on an organic polymer absorbant Porapak Q from sequential samples of air, and recovered for gas chromatography. Peak areas, determined by

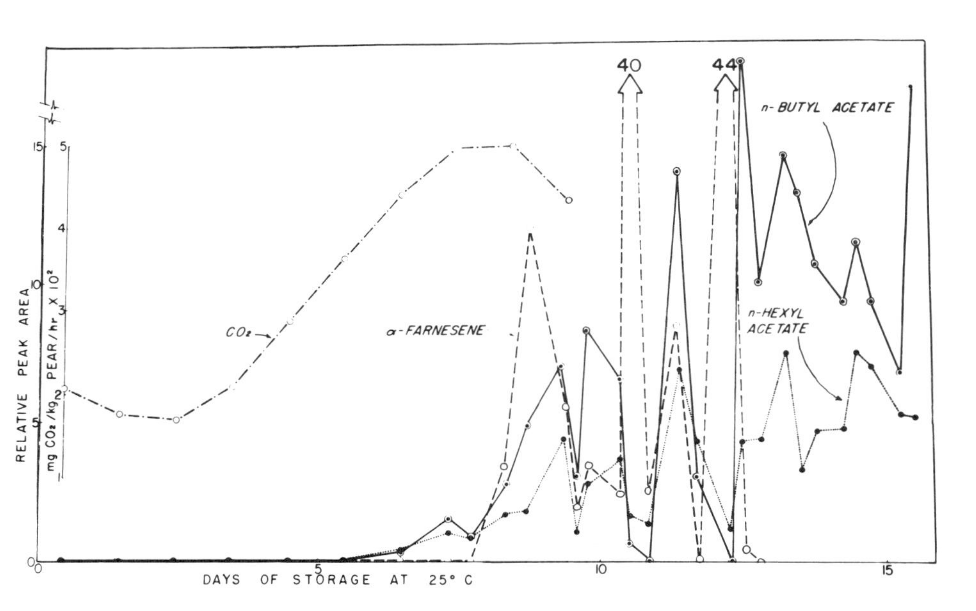

Fig. 2. Rates of individual volatiles from ripening Bartlett pear.

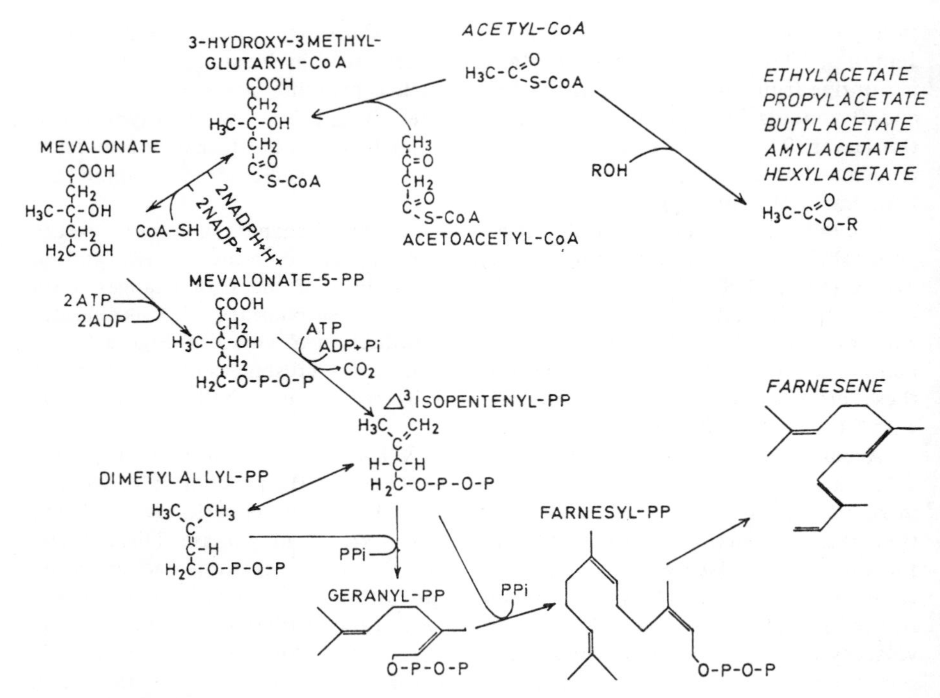

Fig. 3. Possible biosynthetic pathway in Bartlett pear after the respiratory peak.

digital integration and corrected by internal standards, show that the esters *n*-butyl acetate and *n*-hexyl acetate and α-farnesene are produced at rates that vary cyclically. The cycles are out of phase, and some control mechanism seems to alternately activate (or inhibit) the production of terpene and of acetate esters. The biogenesis of volatiles in fruit is a dynamic system.

Figure 3 shows a possible biosynthetic pathway in postclimacteric Bartlett pears. The flavour-bearing decadienoate esters may be explained by β-oxidation of linoleic acid as outlined recently (Fig. 4). Radioactive labelling tests supporting this hypothesis are in progress. Sixty compounds so far have been isolated from Bartlett pears by Jennings & Tressl (1974), 9 acetate esters, 41 C_8 to C_{18} esters of different degrees of unsaturation, and 7 primary alcohols. All volatiles in Bartlett pears may be explained by β-oxidation of saturated or unsaturated acids (Lynen, 1965; Stoffel, 1966). The enzyme complex catalysing β-oxidation is sited in the mitochondria and is inactivated during disintegration of the plant material.

Conversion of mevalonic acid into monoterpenes and sesquiterpenes

According to Bruemmer (1972) terpenoids, constituting a major class of flavour compounds in citrus fruit, are probably synthesized in the juice sac from mevalonic acid via isoprenoid pyrophosphates (Fig. 3). Labelling experiments with [^{14}C]meva-

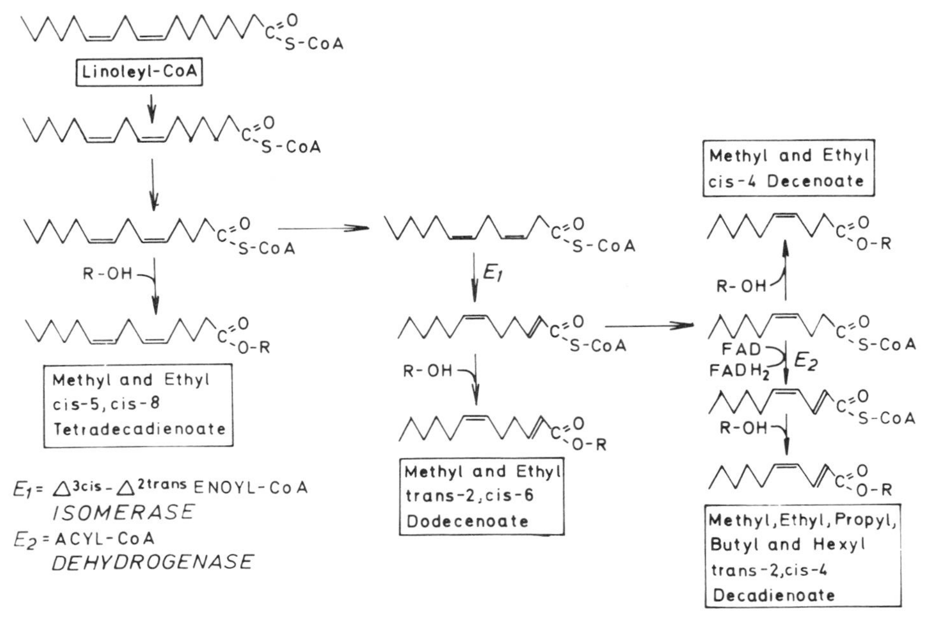

Fig. 4. Reaction scheme for production of unsaturated esters by β-oxidation of linoleic acid.

lonic acid have shown that linaloyl pyrophosphate is the major product and is supposed to be a key intermediate in ring formation of monoterpenes in citrus. In tracer studies with peppermint cuttings, Croteau (1972) showed that the sites of monoterpene and sesquiterpene synthesis were probably compartmented within the oil glands. Incorporation of [^{14}C]mevalonic acid via monoterpenes and sesquiterpenes is shown. The isomerization of geranyl pyrophosphate to neryl pyrophosphate and ring formation as neryl pyrophosphate seem to be key steps in the synthesis of cyclic monoterpene in peppermint. So far little is known about the enzymes catalysing ring formations and rearrangements of terpenes.

Conversion of chain-branched amino acids (L-leucine, L-isoleucine, L-valine) into volatiles

Methyl-branched aliphatic esters, alcohols and acids may be explained by the conversion of L-leucine, L-isoleucine and L-valine (Fig. 1). Many of the major aroma components of banana are branched-chain esters and alcohols (Wick et al., 1969; Tressl et al., 1970). 3-Methylbutyl ester, 3-methylbutanoates and 3-methylbutan-1-ol are derived from L-leucine, the 2-methylpropyl esters, the 2-methylpropanoates and 2-methylpropan-1-ol from L-valine as shown in labelling experiments with postclimacteric banana slices and radioactively labelled precursors. The results and methods have been reported in detail (Tressl et al., 1970); some results of the conversion of [U-^{14}C]L-leucine into volatiles by postclimacteric banana slices are summarized in Table 1. After the climacteric rise in respiration, the mass fraction of free leucine and valine increased from 50 to 150 mg per kg, whereas L-isoleucine and other amino acids remained constant (Drawert et al., 1970). The labelling tests showed that leucine is transformed via 2-oxo-4-methylpentanoic acid into both 3-methylbutyl esters and 3-methylbutanoates. The enzymes involved are sited in the mitochondria. The amino acid is

Table 1a. Conversion of $[U-^{14}C]$L-leucine into volatiles by postclimacteric banana slices.[1]

		Distribution of radioactivity between the volatile components (%)
1	Ethyl acetate	
	3-Methylbutanal	0.5
2	2-Methylbutyl acetate	
	Methyl *3-methylbutanoate*[2]	1.1
3	Ethyl butanoate	0.1
4	*n*-Butyl acetate	
	Ethyl *3-methylbutanoate*	2.4
5	*3-Methylbutyl* acetate	*16.0*[3]
6	*3-Methylbutan-1-ol*	*18.0*
7	2-Methylpropyl *3-methylbutanoate*	
	3-Methylbutyl 2-methylbutanoate	*1.9*
8	*n*-Butyl butanoate	
	2-Pentyl butanoate	10.0
9	*3-Methylbutyl* butanoate	*10.0*
10	*3-Methylbutyl 3-methylbutanoate*	*9.0*
11	*Methyl 2-oxo-4-methylpentanoate*	*4.0*
12	*n*-Hexyl butanoate	
	3-Methylbutyl hexanoate	
	2-Heptyl *3-methylbutanoate*	*25.0*
	4-Hepten-2-yl *3-methylbutanoate*	
13	*3-Methylbutanoic* acid	*2.0*

1. Mass and dimensions of banana discs: 40 g (3 mm x 20 mm); incubation time: 3 h; precursor: $[U-^{14}C]$L-leucine, activity 50 μCi; radioactivity recovered in the aroma extract 0.5%.
2. In italics: part of the molecule where the radioactivity is located.
3. In italics: major radioactive volatiles.

Table 1b. Conversion of $[U-^{14}C]$L-leucine in 0.4 M saccharose solution into volatile components by postclimacteric banana tissue slices.[1]

Distribution of radioactivity between the volatile components after saponification of the aroma extract (%)

Alcohols (47% of total)		*Acids (53% of total)*	
2-Methylbutan-1-ol	0	Acetic acid	0.1
1-Butanol	0	Butanoic acid	1.7
3-Methylbutan-1-ol	76.0	3-Methylbutanoic acid	35.0
2-Heptanol	23.0	Hexanoic acid	1.2
1-Hexanol	1.0	2-Oxo-4-methylpentanoic acid	62.0

1. Mass and dimensions of banana discs: 40 g (3 mm x 20 mm); incubation time: 3 h; precursor: $[U-^{14}C]$L-leucine, activity 50 μCi; radioactivity recovered in the aroma extract 0.5%.

first transaminated to the 3-keto acid which is oxidized and decarboxylated to the corresponding 'activated' aldehyde bound to thiamine pyrophosphate. The key substance for the biosynthesis of 3-methylbutanoates is supposed to be 3-methylbutyryl-CoA derived from the 'activated' aldehyde via lipoic acid. The key substance for the formation of 3-methylbutyl esters, 3-methylbutan-1-ol is formed by enzymic reduction of the aldehyde. The reaction is analogous to the transformation of pyruvate into acetyl-CoA and ethanol, respectively. In apples (Drawert et al., 1972) and strawberries (Tressl et al., 1969), the major branched-chain components are derived from L-isoleucine.

Conversion of L-phenylalanine and tyrosine into aroma substances

The aromatic amino acids phenylalanine and tyrosine are known to be derived from carbohydrates via 3-deoxy-D-arabinoheptulosonate-7-phosphate, shikimic acid, chorismic acid and prephenic acid (Metzner, 1973). Phenylalanine is supposed to be a key intermediate in the biosynthesis of phenyl propane derivates (Fig. 5).

Some of the enzymes involved in the transformation of phenylalanine into phenolic acids have been isolated and characterized by Pridham (1965), Russel & Lonn (1969), Shimida et al. (1970), and Vaughan & Butt (1969). Labelling tests with banana slices and [1-^{14}C]phenylalanine (chain labelled) have shown labelling

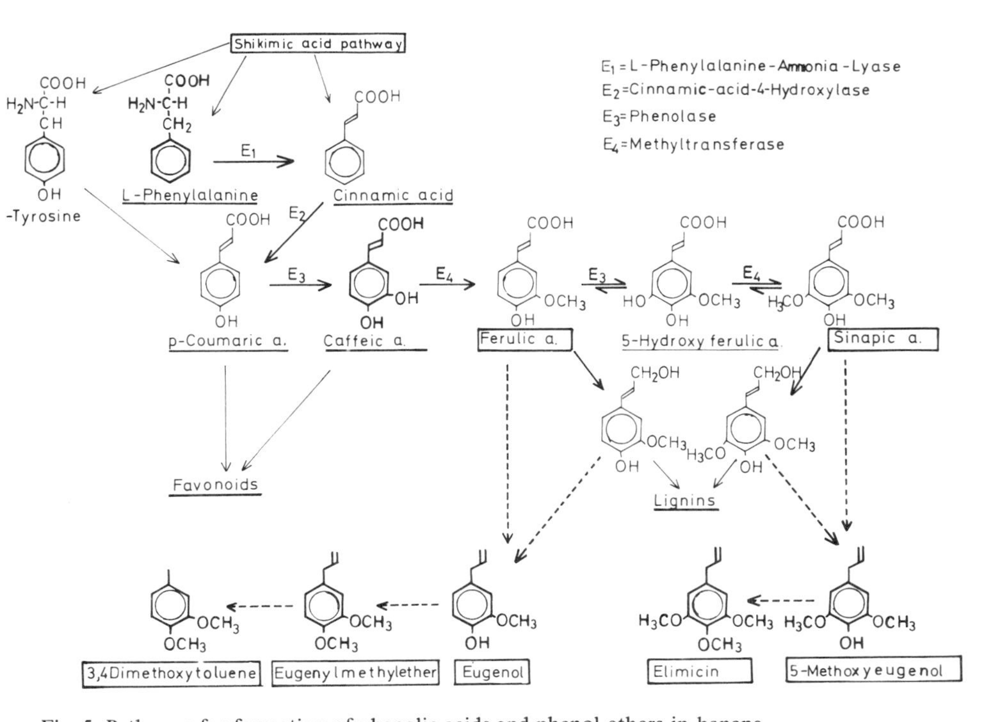

Fig. 5. Pathways for formation of phenolic acids and phenol ethers in banana.

Table 2a. Conversion of [2-^{14}C]L-phenylalanine into aroma components by banana slices.[1]

		Distribution of radioactivity between volatile components (%)
1	2-Phenylethanol	67
2	Eugenol	0
3	Eugenyl methyl ether	0
4	2-Phenethyl acetate	23
5	2-Phenethyl butanoate	10
6	Elimicin	0
7	5-Methoxyeugenol	0

1. Mass and dimensions of banana discs: 50 g (3 mm x 20 mm); incubation time: 5 h; radioactivity recovered in the aroma extract 0.56%.

Table 2b. Conversion of [1-^{14}C]caffeic acid into phenolic ethers by banana slices.[1]

		Distribution of radioactivity between 6 aroma components (%)	
		slices	sucrose solution
1	Peaks eluted before eugenyl methyl ether	6.0	7.0
2	Eugenyl methyl ether	10.0	5.5
3	Eugenol	14.0	53.0
4	Elimicin	12.0	0
5	5-Methoxyeugenol	6.0	10.5
6	Peaks eluted after 5-Methoxyeugenol	52.0	24.0

1. Mass and dimensions of banana discs: 50 g (3 mm x 20 mm); incubation time: 5 h. Radioactivity recovered in the pentane/ether extract of slices: 0.5%, in that of the sucrose solution: 1.0%; radioactivity of unmetabolized caffeic acid in the extract of slices: 0.45%; in that of sucrose solution: 0.90%.

of phenolic ethers with an allyl system. [1-^{14}C]caffeic acid was converted into eugenyl methyl ether, eugenol and elimicin by postclimacteric banana discs (Table 2). [2-^{14}C]phenylalanine is converted into 2-phenylethanol, phenethyl acetate and phenethyl butanoate as shown in Table 2.

In strawberries, the identified methyl and ethyl cinnamates (Tressl et al., 1969) might be derived by the pathway shown in Fig. 5.

Conversion of fatty acids into volatile components

Many of the aliphatic esters, alcohols, acids and carbonyls may be explained by fatty acid metabolism as summarized in Fig. 1. We investigated the conversion of ^{14}C-labelled C_2 to C_{16} fatty acids by fruit slices in the preclimacteric, climacteric

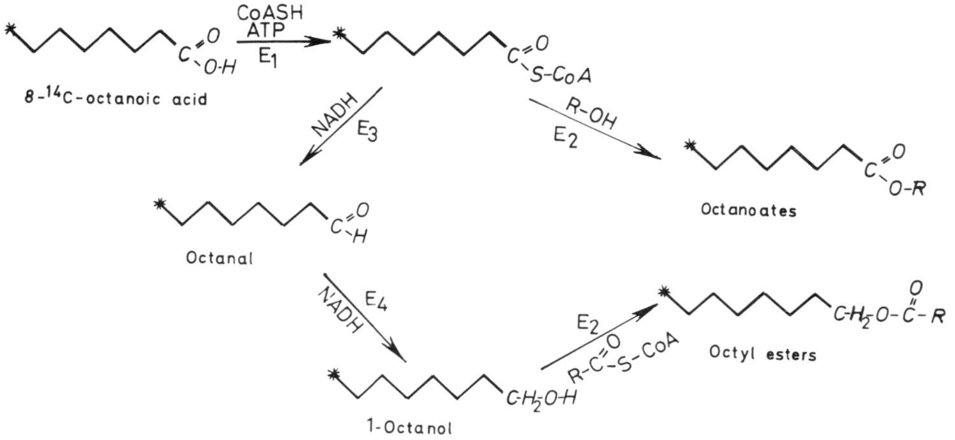

CoASH
ATP
E_1

8-^{14}C-octanoic acid

NADH E_3

R-OH E_2

Octanoates

Octanal

NADH E_4

E_2 R-C(=O)-S-CoA

Octyl esters

1-Octanol

E_1 = Acyl-Thiokinase
E_2 = Acyl-CoA-Alcohol-Transacylase
E_3 = Acyl-CoA-Reductase
E_4 = Alcohol-NAD-Oxydoreductase

Fig. 6. Reaction scheme for conversion of octanoic acid into esters.

and postclimacteric ripening phases (Tressl et al., 1970; Tressl & Drawert, 1971; Tressl et al., 1973; Tressl & Drawert, 1973). Acetate and butanoate were converted into acetate and butanoate esters by postclimacteric discs. Hexanoate, octanoate and decanoate were transformed into esters (Fig. 6). C_6 to C_{10} acids were reduced to the corresponding alcohols, which also undergo transacylation to esters. Besides the acids are metabolized to lower acids by α-oxidation and β-oxidation. [8-^{14}C]octanoate is converted into 1-octanol and 4-hepten-2-ol by climacteric and postclimacteric banana slices, and into 1-octanol and 2-heptanone by postclimacteric strawberry discs. The results and methods have been reported (Tressl et al., 1973) [U-^{14}C]hexadecanoic acid was not converted into volatile constituents by banana or strawberry slices. The results demonstrate that exogeneous C_6 to C_{10} fatty acids are metabolized by climacteric and postclimacteric slices by β-oxidation and α-oxidation, but the conversion into esters could only be demonstrated with postclimacteric tissue slices. This transformation may be explained by a reaction scheme (Fig. 6).

Biogenesis of volatiles in vegetables

If we compare the aroma substances identified in fruit and vegetables, we find some remarkable differences. Apples, pears, bananas, strawberries and peaches release a lot of volatiles during ripening and storage. Esters, alcohols, carbonyls and terpenes are the predominant compounds. When we inhibit the enzymes of a ripe fruit during homogenization, enrich the volatiles by liquid-liquid extraction with pentane and ether and investigate the aroma concentrate by capillary gas chromatography and mass spectrometry, we separate and identify a hundred or two hundred components. These are real metabolites produced by intact tissues at a definite ripening stage.

Thio components and carbonyls are the predominant volatiles in most vegetables (Johnson et al., 1971a, b). If we investigate the volatiles of intact tissues by inhibiting the enzymes during homogenization, we only detect a few fatty acids or other substances with higher boiling temperature. Normally the intact tissues of vegetables only contain non-volatile precursors, which are separated from appropriate enzymes. During homogenization the tissues are destroyed and the non-volatiles

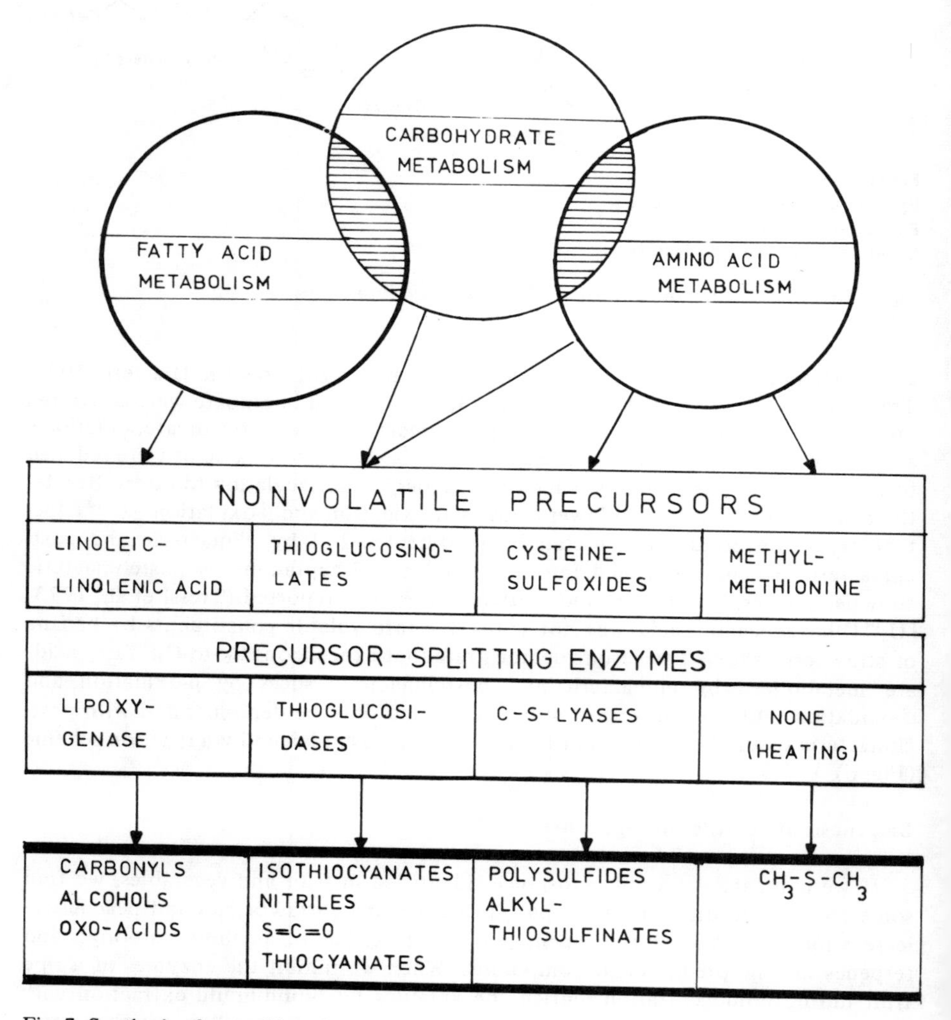

Fig. 7. Synthesis of vegetable volatiles.

are transformed into volatiles. These reactions are catalysed by enzymes (Fig. 7). Sometimes precursors may be split by chemical reactions during the process of heating.

Consequently, the formation of volatiles in fruit and vegetables is quite different. Volatiles in fruit are derived from pathways in the intact tissues during a short phase of ripening. The regulation may be due to plant hormones, or cyclic AMP, but is still unknown. Intact tissues of vegetables synthesize non-volatiles, which are split by appropriate enzymes during crushing of the plant material. The non-volatiles are *S*-alkylcysteinesulphoxides, thioglucosinolates and unsaturated fatty acids, which are transformed into volatile thio components and carbonyls. Of course there are exceptions where typical aroma substances are formed in the intact tissues. The phthalides and selinenes in celery (Gold & Wilson, 1963), 2-methoxy-3-isobutyl pyrazine in bell pepper (Buttery et al., 1969) or 2-isobutyl thiazole in tomatoes (Kazeniac & Hall, 1970) are examples.

Formation of thio compounds

Enzymic splitting of S-alkyl and S-alkenyl-L-cysteinesulphoxides into volatiles

The typical aroma substances of *Allium* species are produced by enzymic splitting of the non-volatile *S*-alkyl and *S*-alkenylcysteinesulphoxides. The responsible

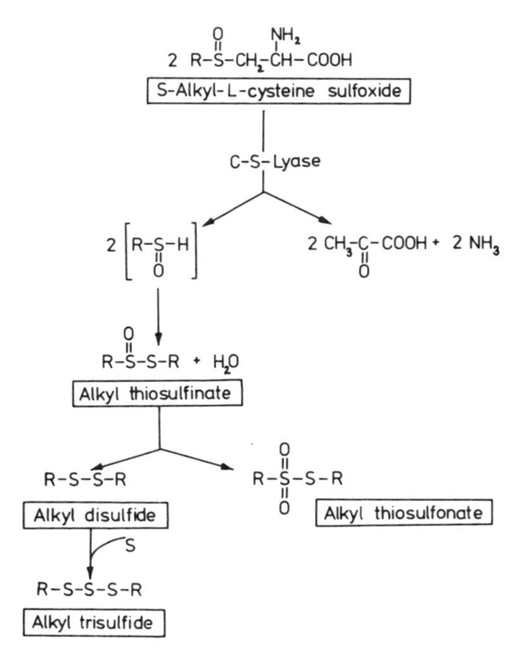

Fig. 8. Enzymic splitting of S-alkyl-L-cysteine sulphoxides in the formation of aroma compounds in *Allium* species.

enzymes are C-S lyases (alliinases) which are separated from their substrates in the intact tissues. The precursors, the enzymes and the products were investigated by Stoll & Seebeck (1949, 1951), Virtanen (1962), Carson (1967), and Schwimmer & Friedman (1972).

Garlic (Allium sativum) *S*-allyl-L-cysteinesulphoxide is the non-volatile precursor of garlic flavour as shown by Stoll & Seebeck (1951). C-S lyases (alliin allylsulphenate lyase, EC 4.4.1.4) catalyse the transformation of the odourless precursor into sulphur-containing volatiles which are responsible for the typical flavour of garlic. The primary sulphur-containing compound is considered to be a sulphenic acid (Fig. 8). Two molecules of the instable sulphenic acid presumably combine to form one molecule of allyl thiosulphinate (allicin), the active odour principle of fresh garlic. The thiosulphinates can undergo disproportionation to allyl disulphide and allyl thiosulphonates, which are the predominant thio components in homogenized garlic. During this chemical reaction allyl trisulphides and allyl tetrasulphides may be formed.

S-methyl-L-cysteine sulphoxide proved to be another precursor in garlic. It is decomposed analogously to dimethyl disulphide, methyl thiosulphonate and mixed allyl methyl polysulphides identified in garlic (Brodnitz & Pascale, 1971; Brodnitz et al., 1971).

Onion (Allium cepa) The situation in onions is more complicated than in garlic. Virtanen (1962), identified *S*-methyl, *S*-propyl and *S*-prop-1-enyl-L-cysteinesulphoxides as aroma precursors. The alkylcysteinesulphoxides are transformed into the corresponding disulphides and thiosulphonates (Fig. 8). The precursor of the lachrymatory factor in onion is *S*-prop-1-enyl-L-cysteinesulphoxide. It decomposes to propenylsulphenic acid, with a half-value time of 90 s. Propenylsulphenic acid can then undergo rearrangement to form the more stable isolatable thiopropanal-*S*-oxide. This component is believed to be the lachrymator. The propenyl-cysteinesulphoxide is further transformed into *cis* and *trans* methyl propenyl disulphides, *cis* and *trans* propyl propenyl disulphides, *cis* and *trans* dipropenyl disulphides and into the corresponding trisulphides (Boelens et al., 1971). The carbonyl compounds identified, propanal, 2-methylpentanal and 2-methylpent-2-enal, may arise from chemical reactions of thiopropanal-*S*-oxide and puryvic acid as postulated by Virtanen (1962).

According to Boelens et al. (1971), the propylthiosulphonates smell strongly and distinctly of freshly cut onions. The propyl and propenyl disulphides and trisulphides taste of cooked onions. 2,4 and 3,4-dimethylthiophenes which are formed during the heating of dipropenyl disulphides are responsible for the flavour of fried onions. The biosynthesis of the propenylcysteinesulphoxide was investigated by Schwimmer & Friedman (1972). According to Schwimmer & Guadagni (1968), the *S*-substituent is derived from the amino acid L-valine via 2-oxo-3-methylbutanoate, 2-methylpropenoate, and *S*-(2-carboxypropyl)-L-cysteine. In ripe onions, part of the flavour precursors may be bound as γ-glutamyl peptides (γ-L-glutamyl-*S*-propenyl cysteinesulphoxides). Transpeptidase and C-S lyases allow flavour production. The corresponding L-cysteinesulphoxide lyases have an optimum at alkaline pH values (Schwimmer & Friedman, 1972). *S*-Methyl-cysteinesulphoxide was also detected in radish root.

Enzymic genesis of volatiles from glucosinolates (thioglucosides)

Almost all vegetables containing glucosinolates belong to the family Cruciferae (Van Etten et al., 1969). Today there are 50 identified glucosinolates of different structure, many characterized by Ettlinger & Kjaer (1968). Glucosinolates are non-volatile flavour precursors which are hydrolysed to volatiles when the plant is crushed. Hydrolysis is catalysed by glucosinolases produced in the plant. Some of the released components are progoitrins and form substances such as 5-vinyloxazolidine-2-thione (goitrin), a strong antithyroid agent (Greer, 1962).

Radish (Raphanus sativus) The radish is widely known both as food and as an old remedy for ailments of the liver and bile. Not surprisingly many attempts have been made to isolate and identify the pungent principle of radish. In 1931 Heiduschka and Zwergel postulated Structure 1 (Fig. 9) for a component which they isolated from 50 kg of fresh radish by distillation. In 1948 Schmid and Karrer demonstrated in radish a thioglucoside that was hydrolysed to an *iso*thiocyanate to which Structure 2 was assigned. Friis & Kjaer (1966) extracted radish roots with

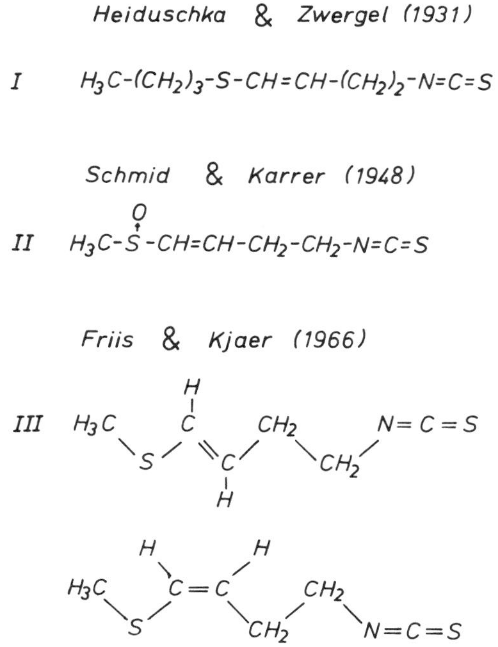

Heiduschka & Zwergel (1931)

I $H_3C\text{-}(CH_2)_3\text{-}S\text{-}CH{=}CH\text{-}(CH_2)_2\text{-}N{=}C{=}S$

Schmid & Karrer (1948)

II $H_3C\text{-}\overset{O}{\overset{\uparrow}{S}}\text{-}CH{=}CH\text{-}CH_2\text{-}CH_2\text{-}N{=}C{=}S$

Friis & Kjaer (1966)

III $H_3C\text{-}S\text{-}CH{=}CH\text{-}CH_2\text{-}CH_2\text{-}N{=}C{=}S$ (trans, H above and below C=C)

$H_3C\text{-}S\text{-}CH{=}CH\text{-}CH_2\text{-}CH_2\text{-}N{=}C{=}S$ (cis, H and H on same side of C=C)

Fig. 9. Proposed chemical structures of the pungent principle in radish root.

methanol and isolated a component, which was purified by gas chromatography. Structure 3 was confirmed by infrared, nuclear magnetic resonance and mass spectrometry. The quotient of *trans* to *cis* isomer was 4:1. According to Friis & Kjaer 4-methylthio-3-butenyl-isothiocyanate, the pungent principle of radish, is released from a non-volatile glucosinolate by glucosinolases in disintegrated plant material.

After inhibition of the enzymes with methanol during homogenization to minimize enzymic reactions, we extracted the volatiles with pentane and methylene chloride and obtained an extract with the characteristic odour of freshly cut radish. When we extracted without enzyme inhibition, the extract lacked the characteristic pungency, but had the unpleasant odour of methanethiol.

Fig. 10 shows a gas chromatogram of the radish extract with enzymic hydrolysis inhibited. The analysis was performed with two selective detectors. In the upper chromatogram, the nitrogen-sensitive AFID detector shows only components containing nitrogen. The sulphur-containing components are recorded with a sulphur-selective detector as described by Brody & Chaney (1966).

Fig. 10 demonstrates that Peaks 1 to 5 contain only sulphur compounds whereas Peaks 6 to 10 contain nitrogen as well as sulphur components. They were enriched and purified by preparative gas chromatography and characterized by mass spectra, the more abundant components also by infrared spectra. The results are shown in

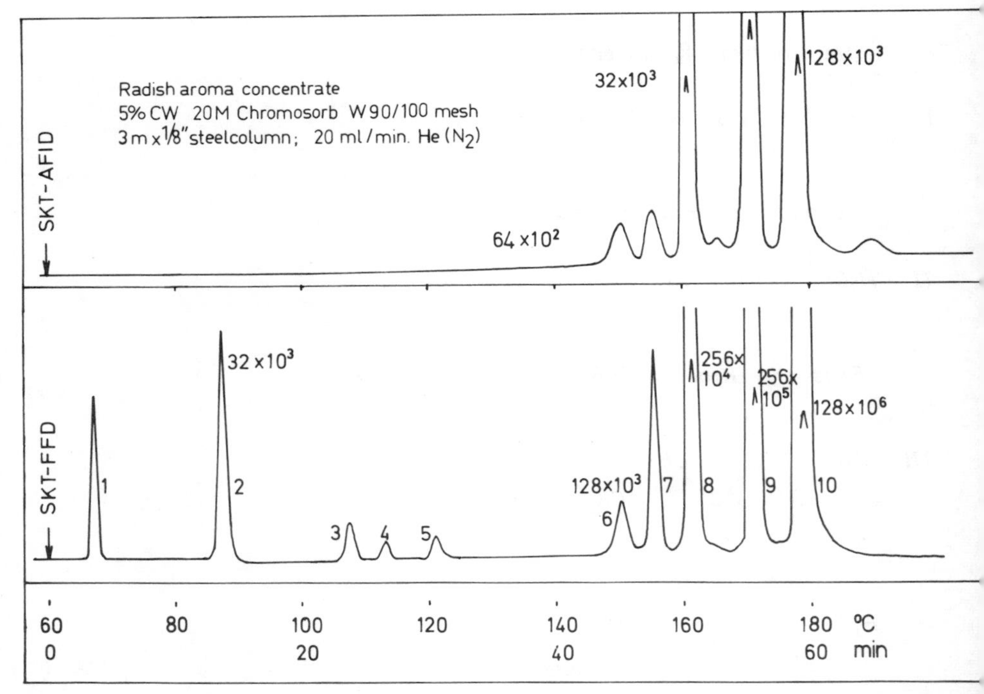

Fig. 10. Gas chromatograms of aroma extract of radish. Upper chromatogram: nitrogen-sensitive detector, lower chromatogram: sulphur-sensitive detector.

Table 3. Formation of volatile thio compounds in radish. Extraction analysis.

	Compound	M	Relative peak area	
			enzyme inhibited	enzyme uninhibited
1	S N=C=S	159	1150	120
2	S N=C=S	159	340	0
3	S N=C=S	161	110	39
4	S N=C=S	173	40	10
5	S N=C=S	147	15	4
6	S C≡N	127	340	5
7	S C≡N	127	10	0
8	S C≡N	129	11	0
9	$CH_3-S-S-S-CH_3$	126	0	120

Table 3. The most abundant component in radish root is *trans*-4-methylthio-3-butenyl *iso*thiocyanate as demonstrated by Friis & Kjaer (1966). The known *cis* isomer and (another) four more methylthioalkyl-substituted *iso*thiocyanates have been identified besides three nitriles (Table 3). According to Friis & Kjaer, the thioglucosinolase reaction in radish causes rapid decomposition of the glucosinolates. Therefore the non-volatile precursors may be partly transformed into *iso*thiocyanates and nitriles in the mature plant. Homogenization without inhibition of the enzymes considerably decreases the amount of *iso*thiocyanates and nitriles. During this decomposition considerable amounts of volatile thio components are formed. Table 3 shows that the methylthioalkenyl ethers are decomposed faster than the methylthioalkyl ethers. According to Bohlmann et al. (1963) methylthiene ethers are unstable and can be transformed into methanethiol and aldehydes. At least five of the glucosinolates are transformed into volatiles when the plant is crushed. The glucosinolates are hydrolysed to unstable aglycones (thiohydroxamic-*O*-sulphonates), which rearrange to *iso*thiocyanates and nitriles. During this reaction the methylthioalkenyl ether components may be transformed into methanethiol and dimethyl disulphide. The functional group of isothiocyanates is hydrolysed to carbonyl sulphide, COS. Fig. 11 shows reactions that could explain the decomposition of the most abundant precursor.

Cabbage (Brassica species) In *Brassica* foods, such as cabbage and cauliflower, *S*-methylcysteinesulphoxide is a known precursor leading to dimethyl disulphide during the crushing and heating of the plant as shown in Tables 4 and 5. Baley et al. (1961) and MacLeod & MacLeod (1970), identified some nitriles and *iso*thiocyanates that might be formed during the disintegration of the plant. We investi-

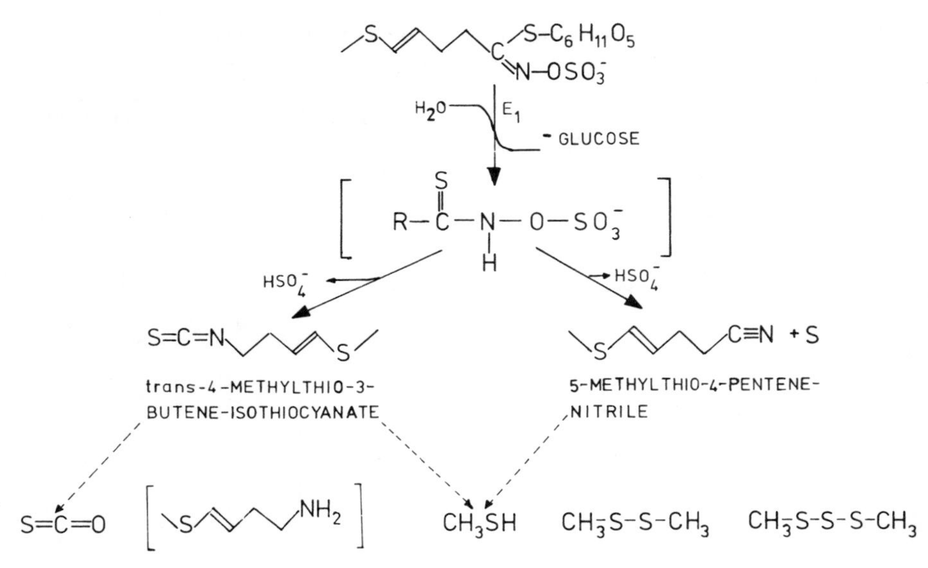

Fig. 11. Possible reactions explaining the formation of volatiles in disintegrated radish root. E = enzyme.

Table 4. Formation of volatile thio compounds in vegetables (mg/kg). Headspace analysis. A: enzyme inhibited; B: enzyme not inhibited, 10 min, room temperature; C: enzyme not inhibited, 10 min, 100 °C.

	S=C=O	CH_3-S-H	CH_3-CH_2-S-H	$CH_3-S-S-CH_3$	CH_3-S-CH_3
Radish					
A	0.4	0.05	0.18	0.2	n.d.
B	360.0	39.0	0.14	80.0	n.d.
C	300.0	87.0	7.00	160.0	8.20
Cabbage (white)					
A	0.7	n.d.	n.d.	0.01	n.d.
B	15.4	n.d.	n.d.	0.28	n.d.
C	8.2	n.d.	0.14	5.0	26.0
Asparagus					
A	n.d.	n.d.	n.d.	n.d.	n.d.
B	n.d.	n.d.	n.d.	n.d.	n.d.
C	n.d.	n.d.	n.d.	n.d.	3.3
Celery					
A	n.d.	n.d.	n.d.	n.d.	n.d.
B	n.d.	n.d.	n.d.	n.d.	0.04
C	0.02	n.d.	n.d.	n.d.	36.0
Tomato					
A	0.03	n.d.	n.d.	0.06	n.d.
B	0.04	n.d.	n.d.	n.d.	n.d.
C	0.04	n.d.	0.04	n.d.	0.46

n.d. = not detectable.

Table 5. Formation of volatile thio compounds in cabbage. Extraction analysis.

	Compound	M	Relative peak area	
			enzyme inhibited	enzyme uninhibited
1	N=C=S	99	46.2	11.0
2	N=C=S	113	27.0	1.0
3	N=C=S	129	2.0	0.0
4	N=C=S	143	2.5	4.0
5	S N=C=S	147	3.0	4.0
6	C≡N	67	11.0	1.0
7	C≡N	81	8.0	1.0
8	S C≡N	115	6.0	6.0
9	$CH_3-S-S-CH_3$	94	0.0	16.0
10	$CH_3-S-S-S-CH_3$	126	0.0	6.0
11	$CH_3-S-S-S-S-CH_3$	158	0.0	3.0
12	$CH_3-S-S-S-CH_2-CH_3$	140	0.0	1.5

gated the formation of volatiles in cabbage as for radish. The results are summarized in Table 5. In the enzyme-inhibited extracts, six *iso*thiocyanates and three nitriles were identified by comparing the mass spectra with published spectra. Only small amounts of carbonyl sulphide and dimethyl sulphide were detected as sulphur-containing components of low boiling temperature. In the uninhibited extracts, the amounts of *iso*thiocyanates and nitriles diminished considerably, whereas carbonyl sulphide and dimethyl sulphide increased. There is a similar hydrolysis of glucosinolates to sulphur-containing volatiles as in radish. As there are no methylthioalkenyl ether precursors in *Brassica,* no methanethiol is formed during the crushing of the plant. Dimethyl sulphide, trimethyl and tetramethyl sulphides and a methyl ethyl trisulphide were identified as shown in Table 5.

Formation of dimethyl sulfide by splitting of methylmethionine-sulphonium ions

Methylmethionine sulphonium salts, first identified by Challenger & Hayward (1954) in asparagus, seem to be common methionine derivates in many vegetables and other plants. During heating the precursor decomposes to dimethyl sulphide in the disintegrated material as shown in Table 4. Dimethyl sulphide has a characteristic flavour quality to a threshold mass fraction of 0.1 to 0.2×10^{-9}. This reaction leads to a principal aroma constituent in asparagus (Ney & Freytag, 1972), in canned tomato juice (Kazeniac & Hall, 1970), and contributes to the flavour of canned or cooked vegetables.

Other sulphur-containing amino acids such as methionine, methylcysteine and cysteine are also transformed into sulphur-containing volatiles during heating. The volatiles are formed in Maillard reactions (Tressl et al., 1975).

In a similar manner methylcysteine and methionine are transformed into methanethiol, dimethyl disulphide, into the Strecker aldehydes, the corresponding alcohols and thioethers. Cysteine and cystine are transformed into H_2S and acetaldehyde by Strecker degradation forming thiophenes, thiazoles and other sulphur-containing heterocyclic compounds at higher temperatures. The Strecker aldehyde of methionine (methional) is known to be the principal aroma constituent of cooked potatoes.

Formation of carbonyls and alcohols by enzymic splitting of unsaturated fatty acids

Linoleic and linolenic acid are non-volatile potent flavour precursors in vegetables and fruit. During crushing of the plant, the polyenoic acids, possessing a *cis, cis*-pantadiene system, are transformed via 9 or 13-hydroperoxy-octadecadienoic (or trienoic) acid into saturated and unsaturated carbonyls, alcohols and oxo acids. The reaction needs oxygen and is catalysed by a lipoxygenase system (EC 1.13.11.12).

Cucumber Investigating the enzyme-inhibited, methylated aroma extract of cucumbers by gas chromatography and mass spectrometry, we have only identified fatty acids. The predominant components are hexadecanoic, oleic, linoleic and linolenic acids. The concentrate has no characteristic flavour. Assay without inhibition of the enzymes yields an aroma extract with a strong characteristic flavour of freshly cut cucumbers. Gas chromatography and mass spectrometry of the non-enzyme-inhibited extract show considerable decrease in the amounts of linoleic and linolenic acid while 6 to 10 components of lower boiling temperature appear in the chromatogram. Some of the results are summarized in Table 6. *trans*-2,*cis*-6-Nonadienal, the flavour bearing component of cucumbers was first identified by

Table 6. Formation of cucumber volatiles (mg/kg). Extraction analysis.

	Compound	Enzyme inhibited	Enzyme uninhibited	Enzyme uninhibited, linoleic ester added[1]	Enzyme uninhibited, linolenic ester added[1]
1	Hexanal	0	1.5	25.0	0.2
2	*trans*-2-Hexenal	0	1.8	0.2	8.5
3	*trans*-2-Nonenal	0	0.8	5.6	0.1
4	*trans*-2,*cis*-6-Nonadienal	0	2.5	1.1	6.2
5	9-Oxononanoic acid	0	0	17.5	15.2
6	12-Oxododecenoic acid	0	0	8.2	4.3

1. Linolenic and linoleic acid methyl ester, 0.1 ml. Homogenized with tissue for 10 min at room temperature.

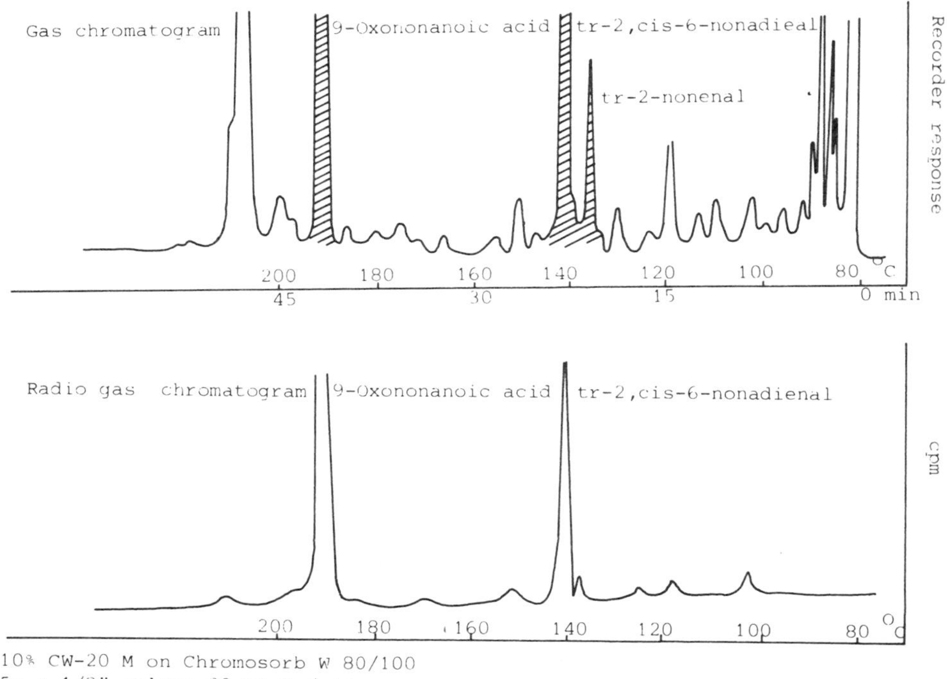

Fig. 12. Enzymatic splitting of [U-^{14}C]linolenic acid into volatiles and oxoacids in unripe banana tissue.

Takei & Ono (1939). *trans*-2-Nonenal, hexanal and *trans*-2-hexenal were identified by Forss et al. (1962). Fleming et al. (1968) showed that the aldehydes were formed by enzymic reactions during crushing of the plant.

Aroma extracts prepared without enzyme inhibition from unripe bananas have a note like cucumbers. In an analogous reaction, linoleic and linolenic acids are transformed into 2-nonenal, 2,6-nonadienal and 9-oxononanoic acid. In ripe bananas, the precursors are split into hexanal, 2-hexenal, and 12-oxododecenoic acid, respectively. We investigated the decomposition of unsaturated fatty acids with radioactively labelled [U-^{14}C]linoleic and [U-^{14}C]linolenic acid and determined the distribution of labelled components by radioactive gas chromatography. Linoleic acid was confirmed as precursor of hexanal, *trans*-2-nonenal, 9-oxononanoic acid and 12-oxododecenoic acid. In homogenates, the aldehydes are partly reduced to the corresponding alcohols while the oxo acids can undergo polymerization or polycondensation forming non-volatiles.

Linolenic acid could be confirmed as precursor of *trans*-2 and *cis*-3-hexenal, *trans*-2, *cis*-6-nonadienal, 9-oxononanoic acid and 12-oxododecenoic acid. Fig. 12 shows the results of the transformation of [U-^{14}C]linolenic acid into volatiles by unripe bananas discs. Fig. 13 shows a possible reaction scheme which may explain

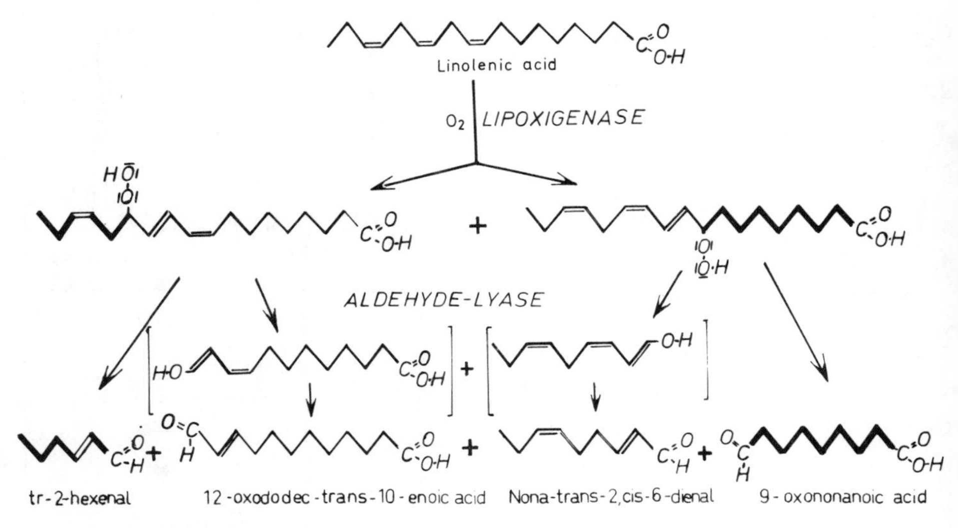

Fig. 13. Reaction scheme which may explain the formation of 2,6-nonadienal in cucumbers. Enzymic splitting of [U-^{14}C]linolenic acid into volatiles and oxoacids.

the transformation of unsaturated acids into volatiles by tissues of cucumbers, tomatoes, bananas and other fruit.

The aldehydes formed are very potent flavour components with characteristic notes and low thresholds (Teranishi et al., 1971). According to Kazeniac & Hall (1970), hexanal, *trans*-2-hexenal, *cis*-3-hexenal and *iso*butyl thiazole, three of more than 200 identified components, contribute most to the typical flavour of fresh tomatoes. When the amounts of these compounds decreased below the threshold for sensory detection, the alcohols and other components became the predominant flavour components. When the amounts of the C_6 aldehydes decreased, the flavour became 'processed'. 'Heated' or 'cooked' tomato flavour was associated with large amounts of dimethyl sulphide, furfural, methional, acetaldehyde, phenylacetaldehyde and 2,4-heptadienal. In asparagus, linoleic acid is transformed into pentanal, hexanal, 2-nonenal, 9-oxononanoic acid, 12-oxododecenoic acid (Tressl, unpublished results).

In other vegetables such as beans or peas, linoleic and linolenic acid are split in a different way (Grosch & Schwenke, 1969; Leu, 1974). Besides hexanal and 2-hexenal a lot of saturated aldehydes, enals and dienals are formed by a lipoxygenase reaction (EC 1.13.11.12) leading to off-flavour components during storage of the frozen vegetables.

In all vegetables, methyl-branched aldehydes are derived from valine, isoleucine and leucine during heating by Strecker degradation as shown in canned tomatoes, asparagus and other vegetables.

References

Baily, S. D., M. L. Bazinet, J. L. Drisoll & A. J. McCarthy, 1961. J. Food Sci. 26: 163.
Biale, J. B., 1964. Science 146: 880.
Boekh, J., 1974. Int. Symp. Geruchs- und Geschmacksstoffe, Bad Pyrmont, 1974.
Boelens, M., P. J. de Valois, H. J. Wobben & A. van der Gen, 1971. J. agric. Food Chem. 19: 984.
Bohlmann, F., P. Herbst & J. Dohrmann, 1963. Chem. Ber. 96: 226.
Brodnitz, M. H. & J. V. Pascale, 1971. J. agric. Food Chem. 19: 269.
Brodnitz, M. H., J. V. Pascale & L. V. Derslice, 1971. J. agric. Food Chem. 19: 273.
Brody, S. S. & J. Chaney, 1966. J. Gas Chromatogr. 4: 42.
Bruemmer, J. H., 1972. Meet. American Chemical Society, Analytical Food Chemistry Division, New York.
Buttery, R. G., R. M. Seifert, D. G. Guadagni & L. C. Ling, 1969. J. agric. Food Chem. 17: 1322.
Carson, J. F., 1967. In: H. W. Schultz (ed.), The Chemistry and Physiology of Flavors, AVI Publishing Company, Inc., Westport, Connecticut, p. 390.
Challenger, F. & B. J. Hayward, 1954. Chem. Ind. 25: 729.
Croteau, R., 1972. Meet. American Chemical Society, Analytical Food Chemistry Division, New York.
Drawert, F., W. Heimann, R. Emberger & R. Tressl, 1972. Chem. Mikrobiol. Technol. Lebensm. 1: 201.
Drawert, F., K. Rolle, W. Heimann, R. Emberger & R. Tressl, 1970. Z. LebensmUnters. Forsch. 144 : 237
Ettlinger, M. G. & A. Kjaer, 1968. Rec. Adv. Phytochem. 1: 59.
Fleming, H. P., W. Y. Cobb, J. L. Etchells & T. A. Bell, 1968. J. Food Sci. 33: 572.
Forss, D. A., E. A. Dunstone, E. H. Ramshaw & W. Stark, 1962. J. Food Sci. 27: 90.
Friis, P. & A. Kjaer, 1966. Acta chem. scand. 20: 698.
Gold, H. J. & C. W. Wilson, 1963. J. Food Sci. 28: 484.
Greer, M. A., 1962. Recent Progr. Horm. Res. 18: 187.
Grosch, W. & D. Schwenke, 1969. LebensmWiss. Technol. 2: 109.
Hansen, E., 1969. Ann. Rev. Plant Physiol. 17: 459.
Heiduschka, A. & J. Zwergel, 1931. J. prakt. Chem. 132: 201.
Heinz, D. E., R. K. Creveling & W. G. Jennings, 1965. J. Food Sci. 30: 641.
Jennings, W. G. & R. Tressl, 1974. Chem. Mikrobiol. Technol. Lebensm. 3: 52.
Johnson, A. E., H. E. Nursten & A. A. Williams, 1971a. Chem. Ind. London: 556.
Johnson, A. E., H. E. Nursten & A. A. Williams, 1971b. Chem. Ind. London: 1212.
Kazeniac, S. J. & R. M. Hall, 1970. J. Food Sci. 35: 519.
Kolattukudy, P. E., 1970. Lipids 5: 259.
Leu, K., 1974. LebensmWiss. Technol. 7:98.
Liebermann, M., K. Kunishi, L. W. Mapson & D. A. Wardale, 1966. Plant Physiol. 41: 376.
Lynen, F., 1965. Angew. Chem. 21: 929.
MacLeod, A. J. & G. MacLeod, 1970. J. Food Sci. 35: 739.
Metzner, H., 1973. Biochemie der Pflanzen. Enke Verlag, Stuttgart.
Ney, K. H. & W. Freytag, 1972. Z. LebensmittelUnters. Forsch. 149: 154.
Pridham, J. B., 1965. Ann. Rev. Plant Physiol. 16: 13.
Romani, R. J. & L. Ku, 1966. J. Food Sci. 31: 558.
Russel, D. W. & E. E. Lonn, 1969. J. Biochem. 113: 109.
Schmid, H. & P. Karrer, 1948. Helv. chim. Acta 31: 1017.
Schwimmer, S. & D. G. Guadagni, 1968. J. Food Sci. 33: 193.
Schwimmer, S. & M. Friedman, 1972. Flavour Ind.: 137.
Shimida, M., T. Yamazaki & T. Higuchi, 1970. Phytochemistry 9: 1.
Stoffel, W., 1966. Naturwissenschaften 53: 62.
Stoll, A. & E. Seebeck, 1949. Helv. chim. Acta 32: 197.
Stoll, A. & E. Seebeck, 1951. Adv. Enzymol. 11: 377.
Straten, S. van & F. de Vrijer, 1973. TNO Report R 4030, Central Institute for Nutrition and Food Research TNO, Zeist, the Netherlands.

Takei, S. & N. Ono, 1939. J. Agric. Chem. Soc. Japan 15: 193.
Teranishi, R., J. Hornstein, P. Issenberg & E. L. Wick, 1971. In: O. R. Fennema (ed.), Flavor Research. Marcel Dekker, Inc., New York.
Tressl, R., F. Drawert & W. Heimann, 1969. Z. Naturforsch. 246: 1201.
Tressl, R., R. Emberger, F. Drawert & W. Heimann, 1970. Z. Naturforsch. 24b: 893.
Tressl, R. & F. Drawert, 1971. Z. Naturforsch. 26b: 774.
Tressl, R. & W. G. Jennings, 1972. J. agric. Food Chem. 20: 189.
Tressl, R., F. Drawert & U. Prenzel, 1973. Chromatographia 6: 7.
Tressl, R. & F. Drawert, 1973. J. agric. Food Chem. 21: 560.
Tressl, R., T. Kossa & M. Holzer, 1975. Chem. Mikrobiol. Technol. Lebensm. In press.
Van Etten, C. H., M. E. Daxenbichler & J. A. Wolff, 1969. J. agric. Food Chem. 17: 483.
Vaughan, P. F. T. & V. S. Butt, 1969. Biochem. J. 113: 109.
Virtanen, A. J., 1962. Angew. Chem. 74: 374.
Wick, E. L., T. Yamanishi, A. Kobayashi, S. Valenzuela & P. Issenberg, 1969. J. agric. Food Chem. 17: 751.

Proc. int. Symp. Aroma Research, Zeist, 1975. Pudoc, Wageningen.

Formation of odorous compounds from hydrogen sulphide and methanethiol, and unsaturated carbonyls

H. T. Badings[1], H. Maarse[2], R. J. C. Kleipool[2], A. C. Tas[2], R. Neeter[1], M. C. ten Noever de Brauw[2]

1. Netherlands Institute for Dairy Research, NIZO, P.O. Box 20, Ede, the Netherlands
2. Central Institute for Nutrition and Food Research TNO, Utrechtseweg 48, Zeist, the Netherlands

Abstract

Strongly odorous compounds may be formed upon addition of H_2S and CH_3SH to unsaturated carbonyls. A well known example is the formation of catty odour from mesityl oxide and hydrogen sulphide.

For the reaction products of 2-alkenals and an alkenone with hydrogen sulphide and methanethiol, characteristics of the flavour, its threshold values, and physico-chemical characteristics were determined.

Under certain conditions, such additions may occur during processing and storage of foods.

Introduction

Sulphur-containing substances play a dominant role in the flavour of many food products, as demonstrated again in a review by Schutte (1974). Many of these substances possess characteristic odours and their flavour thresholds are mostly low. The interest of the flavour industry in the use of volatile sulphur compounds to create and improve flavours is evident from recent patent applications, *inter alia* from Firmenich (1972), Givaudan (1972), Naarden (Boelens et al., 1975) and Polak's Frutal Works (1967).

The formation of odorous compounds upon addition of hydrogen sulphide or thiols to the double bonds of unsaturated compounds has attracted attention since the occasional occurrence of catty taints (= *Ribes* taints) in certain food products (Badings, 1967; Pearce et al., 1967). This odour was caused by 2-mercapto-2-methylpentan-4-one, formed from mesityl oxide and hydrogen sulphide. In air pollution, the same off-odours have been observed to result from similar reactions (Maarse & ten Noever de Brauw, 1972, 1974). The addition reactions occurred readily under mild conditions, even in dilute aqueous media.

Since many food products contain unsaturated carbonyl compounds and volatile sulphur compounds (van Straten & de Vrijer, 1973), such as hydrogen sulphide and methyl mercaptan, we tested whether reactions between such components would occur during processing and storage. Some reactions were performed and the reaction products were analysed by gas chromatography and spectrometry. Subsequently, we investigated whether the compounds formed from their constituents in several model reactions.

We will concentrate here on the reaction of hydrogen sulphide with 2-alkenals, especially *trans*-2-butenal, on account of the interesting results obtained.

Experimental: formation of addition products

1. Reactions of methanethiol with 2-alkenals and 1-octen-3-one

3-Methylthioalkanals and 1-methylthiooctan-3-one were prepared as follows. An ice-cooled solution of the 2-alkenal or 1-octen-3-one in chloroform or benzene was saturated with methanethiol. The reaction was catalysed by a few drops of triethylamine. After reaction for two hours at room temperature, the solvent was removed and the crude product was purified by distillation under reduced pressure.

2. Reactions of H_2S and 2-alkenals in aprotic solvents

The method described in a Givaudan patent (Givaudan, 1972) was followed. First triethylamine (10 ml) and then 52 g 2-butenal were added dropwise to chloroform for 10 and 120 min, respectively, while H_2S was continuously led through the solution, which was cooled to −10 °C. For another 2 h, H_2S was led through the cooled solution, which was then left in the cold for 12 h.

The solution was stirred at room temperature with 125 ml HCl of substance concentration 2 mol·litre^{-1} ; the organic layer was separated, washed again with 125 ml of the HCl and then with water until neutral. The chloroform solution was dried over anhydrous Na_2SO_4, the solvent was removed and the product distilled.

The fraction with a boiling temperature of 58–60°C/pressure 10 mmHg (yield 44 g) dimerized on standing. Another fraction with a higher boiling temperature (100–102 °C/3 mmHg (7 g)) was also obtained. The same procedure was used for preparation of 3-mercaptohexanal, 3-mercaptoheptanal and 3-mercaptononanal.

3. Catalysed reaction between H_2S and 2-butenal

Gaseous H_2S was led through 27 g 2-butenal to which 0.32 ml 40% Triton B was added as a catalyst (Gershbein & Hurd, 1947). During the reaction, the temperature rose to 60 °C, and the viscosity of the mixture increased slowly. The reaction was terminated when the mixture became difficult to stir. The reaction mixture was then distilled under vacuum. Two main fractions were obtained, of which the first one (120–128 °C/12 mmHg (3.8 g)) was analysed and the second one was difficult to investigate, because it polymerized rapidly.

4. Formation of addition products in dilute aqueous solution

A 120-ml ampoule was filled with 100 ml of a solution containing 100 mg/l of the unsaturated substance and an excess (five equivalents) of methanethiol or a mixture of NaSH and H_2S. The mixture was formed from Na_2S dissolved in boiled double-distilled water by adding dilute H_2SO_4 to a pH of about 7. This solution was directly transferred to the ampoule. After flushing out air from the ampoule with a stream of nitrogen gas, the ampoule was closed and left for 24 h at room temperature.

The reaction products were collected by extraction of the aqueous solution with three portions of 2 ml pentane in ether, 1:2 by volume. The extracts were concentrated slowly by careful evaporation of the solvent in a micro-distiller as described by Moinas (1973).

5. Reactions between 2-butenal and H_2S at different pH-values

a. 9 g Na_2S was dissolved in 400 ml boiled demineralized water. This solution was acidified to pH 7 with H_2SO_4 of substance concentration 2 mol·litre^{-1}, after which 0.5 ml 2-butenal, dissolved in 30 ml boiled water, was added. The reaction mixture was stirred for ½ h and left overnight.

Similar experiments were conducted in parallel, with Na_2S solutions acidified to either pH 6, 4.8 or 4. The reaction products were isolated from the aqueous solution by extraction with 50 ml dichloromethane. The extract was dried over anhydrous Na_2SO_4 and the solvent was removed by evaporation.

b. H_2S was passed through 1.5 litre water at room temperature until saturated and 20 g 2-butenal as 250 ml aqueous solution was added dropwise in 2 h. The reaction mixture was subsequently stirred for another 1½ h while H_2S was led through continuously.

The reaction mixture was extracted twice with 70 ml chloroform; the extract was dried over Na_2SO_4 and the solvent was removed by evaporation. Upon distillation, a main fraction (70–75 °C/0.4 mmHg (4.5 g)) and two other fractions, fraction A (75–95 °C/3 mmHg (1 g)) and fraction B (95–120 °C/3 mmHg (0.5 g)), were obtained.

Experimental: methods of analysis

Gas chromatography

A Hewlett Packard model 5750 or a Varian 2740 gas chromatograph equipped with dual flame ionization detectors was used, fitted with glass capillary columns 32 or 65 m long and 0.8 mm internal diam. coated with Carbowax 20 M or SE 30 (Badings, 1975). The temperature of the injection port was limited to 150 °C to prevent decomposition of heat-labile compounds. Analysis was by temperature-programmed gas chromatography (70–130 °C).

For preparative purposes a stainless steel column 2 m long and 4 mm internal diam. was used, packed with 20% SE 30 on Chromosorb W-AW, 60–80 mesh.

The reaction mixtures were also analysed at 110 °C with a glass column 2 m long and 3 mm intern. diam. packed with 10% SE 30 on Chromosorb W-AW, 60–80 mesh. This column was fitted in a Mikrotek M.T. 120 gas chromatograph equipped with a flame photometric S-detector.

Synthesized compounds and extracts from reaction mixtures were usually analysed by direct on-column injection.

Spectral analysis

Mass spectra were determined with either a CH4-Varian MAT mass spectrometer in a GC/MS computer system or a CH5-Varian MAT mass spectrometer in a GC/MS system. Glass capillary columns were used for gas chromatography in both systems; connexion to the CH4 was effected by way of a silicone-membrane separator, and to the CH5 by an all-glass capillary inlet splitter. High-resolution mass spectra of some products were measured on a double-focusing mass spectrometer (731, Varian MAT) with the direct insertion probe.

Infrared spectra were recorded on a Hilger & Watts H-1200 or on a Perkin Elmer 257 spectrophotometer.

Nuclear magnetic resonance spectra were measured at 60 MHz on a Jeol C-60 H spectrometer.

Organoleptic evaluation of addition products

A taste panel of six members experienced in the evaluation of odorous fractions participated in the experiments. The panel members were requested to describe the characteristic flavour of the synthesized products and also to determine the minimum concentration at which they were detectable. For this purpose, each compound was presented in a series of samples with increasing concentrations. The addition products formed under different conditions were also evaluated. Sometimes this was done by smelling the odour at the column outlet during a gas-chromatographic separation of the extract of a reaction mixture.

The flavour thresholds were calculated by the method of Patton & Josephson (1957).

Results

Methanethiol with 2-alkenals and 1-octen-3-one

Synthesis (Experimental 1)

The reaction products, 3-methylthioalkanals and 1-methylthiooctan-3-one, were fairly stable and of acceptable purity.

Reactions in aqueous solution (Experimental 4)

The products formed in aqueous solutions were as expected from the reaction of unsaturated carbonyls with CH_3SH. The results are summarized in Table 1.

H_2S with 2-alkenals

The results of all the analysed reactions between hydrogen sulphide and unsaturated carbonyl compounds were complicated. The experiments showed that numerous products are formed. We therefore extensively studied the reaction products of *trans*-2-butenal and hydrogen sulphide under various conditions.

Table 1. Products of the addition reactions between unsaturated carbonyls and CH_3SH.

Carbonyl	Addition product
2-butenal	3-methylthiobutanal
2-hexenal	3-methylthiohexanal
2-heptenal	3-methylthioheptanal
2-nonenal	3-methylthiononanal
1-octen-3-one	1-methylthiooctan-3-one

2-Butenal and H_2S in aprotic solution (Experimental 2 and 3)

The reaction between hydrogen sulphide and *trans*-2-butenal with triethylamine as a catalyst was carried out according to the directions described in a patent (Givaudan, 1972).

Distillation of the reaction mixture gave as main product (70%), the simple addition product 3-mercaptobutanal (I) (63–65 °C/15 mmHg). On standing, this compound changed into a viscous oil of Structure II as shown by spectral analysis and the formation of a diacetate (Kleipool et al., 1975). The dimerization reaction could be followed easily by the changes in the infrared spectrum.

The monomer (I) could be re-obtained by distillation of II.

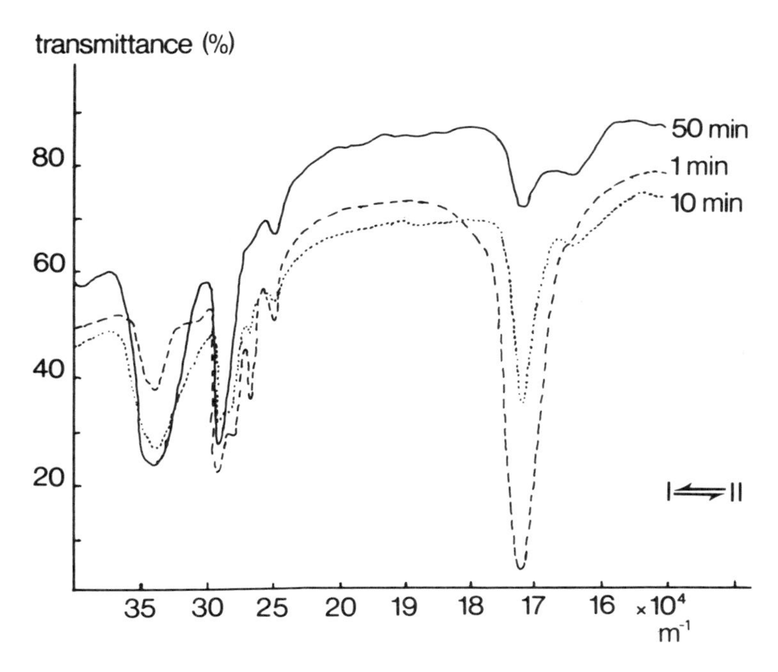

Fig. 1. Changes in the infrared spectrum of 3-mercaptobutanal on standing.

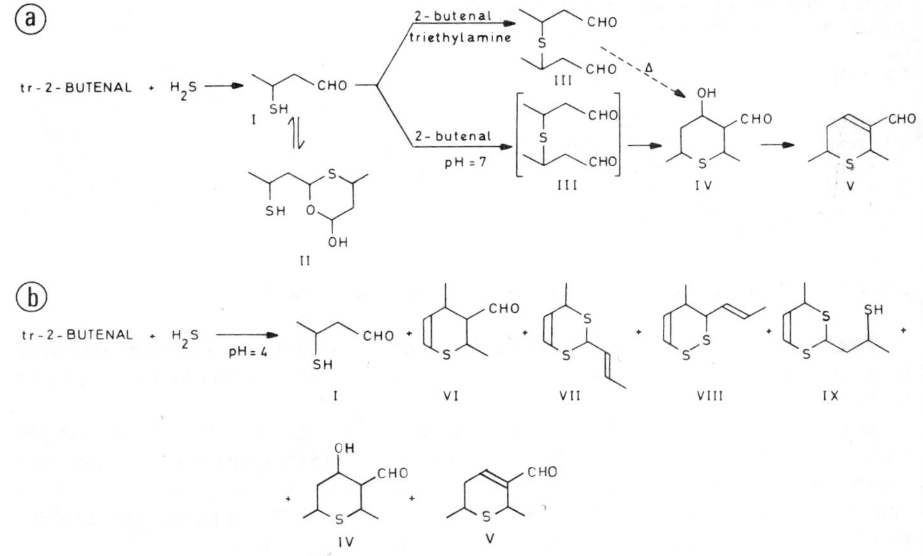

Fig. 2. a. Scheme for the reaction of 2-butenal and H_2S in aqueous solution at pH 7, and in solvents without protons with triethylamine as catalyst. b. Reaction products from 2-butenal and H_2S at pH 4.

The compound 4-methylthietan-2-ol, described in the patent (Givaudan, 1972) was not present in the reaction mixture.

A second fraction, of higher boiling temperature was demonstrated by spectral analysis to have Structure III (Fig. 2) (Kleipool et al., 1975). On heat treatment, e.g. during gas chromatography at a temperature higher than 130 °C, this compound gave two other products, isolated and identified as: 2,6-dimethyl-4-hydroxy-3-formylthiane, IV, and 5,6-dihydro-2,6-dimethyl-3-formyl-(2*H*)-thiin, V. Product IV may be considered as the result of an intramolecular aldol condensation; Product V originates from the loss of a molecule of water from a molecule of Product IV (Fig. 2).

Products III and V had already been found by Asinger & Fisher (1967). When H_2S was led into 2-butenal with Triton B as a catalyst, as described by Gershbein & Hurd (1947), Products IV and V were obtained as the main components. Contrary to Gershbein & Hurd (1947), Compound III could not be isolated.

2-Butenal and H_2S in aqueous solution (Experimental 4 and 5)

Neutral solution When the reaction was in neutral aqueous solution, essentially the same components were formed (Fig. 2). Only the dialdehyde (III) could not be detected, presumable because it spontaneously underwent the aldol condensation.

Table 2. Reaction products from 2-butenal and H_2S.

Peak no.	Identity[1]	Mol. weight	Reaction type[2]	Evidence	
				Kovats index SE 30; 110 °C	spectral
1	I	104	A–F	840	i.r., m.s.
2	unknown		A	940	
3	unknown	118	A–D	979	m.s.
4	unknown		A, C	1092	
5	unknown		A, C, E	1140	
6	unknown		A	1158	
7	unknown		A	1182	
8	VI	156	A, D	1210	n.m.r., i.r., m.s.
9a	unknown	152	A, C, D, E	1225	m.s.
9b	unknown	154	D	1227	m.s.
10	V	156	B, C, E	1240	n.m.r., i.r., m.s.
11	III	174	F	1288	n.m.r., i.r., m.s.
12	VIII	172	A, D	1322	n.m.r., i.r., m.s.
13	IV	174	B, C, E	1324	n.m.r., i.r., m.s.
14	XII	172	A, D	1331	n.m.r., i.r., m.s.
15	IV	174	B, C, E	1342	n.m.r., i.r., m.s.
16	VIII	172	A, D	1344	n.m.r., i.r., m.s.
17	unknown		C, E	1369	
18	unknown		B, C, E	1397	
19	unknown	174?	D	1410	m.s.
20	unknown		C, D	1415	
21	unknown		D	1430	
22	unknown	190	A–D	1482	m.s.
23	IX	206	A, D	1530	n.m.r., i.r., m.s.

1. Numbers correspond to those in Figure 2.
2. Type A; reaction in aqueous solution, Experimental 5b.
   Type B; catalysed reaction without solvent, Experimental 3.
   Type C; reaction in aqueous solution in ampoule, Experimental 4.
   Type D; reaction in aqueous solution at pH 4, Experimental 5a.
   Type E; reaction in aqueous solution at pH 7, Experimental 5a.
   Type F; reaction in an aprotic solvent, catalysed by triethylamine, Experimental 2.

Acidic solution In acid solution, the reaction was much more complicated. The main component was identified as 2-*trans*-prop-enyl-4-methyl-(4*H*)-1,3-dithiin (VII). Many difficulties were encountered in isolating this substance pure, since it easily rearranged into 3,4-dihydro-4-methyl-3-*trans*-propenyl-1,2-dithiin (VIII), when heated. Only with mild conditions of gas chromatography ($\leq$ 110 °C) could rearrangement largely be prevented.

Compounds VII and VIII both occur in two diastereoisomeric forms; the two isomers of VIII were easily separated by gas chromatography (Table 2, Peaks 12 and 16). The rearrangement reaction could be demonstrated by comparing the partial n.m.r. spectra (Fig. 3) of Compounds VII and VIII as well as of their mixtures obtained by gas chromatography of VII under different conditions. At 100 °C, on glass or metal columns, only a small amount of VIII was formed but at 150 °C, the mixture consisted mainly of VIII.

Table 2 shows that the identity of many reaction products is still unknown, especially those of lower building temperatures. Some of these compounds are unstable and decompose or rearrange during analysis.

2-(2′-Mercaptopropyl)-4-methyl-(4*H*)-1,3-dithiin (IX, Fig. 2; Table 2) was found in the higher boiling fraction (Fraction B, experimental 5b).

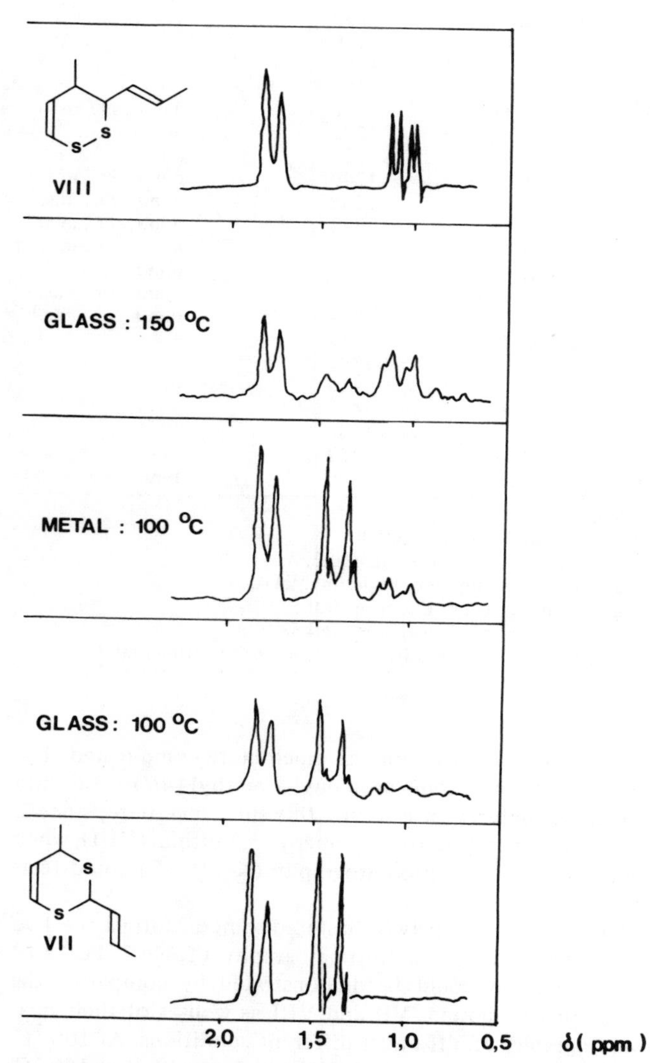

Fig. 3. Partial nuclear magnetic resonance spectra showing the rearrangement of VII to VIII (Fig. 2) as a function of temperature and column material in the gas chromatograph.

Organoleptic evaluations

The results of the flavour evaluation are given in Table 3. The nature of the odour and the threshold are not always those of pure compounds. Often mixtures of different compounds and of isomers of one compound are present in the reaction mixtures. But they are meaningful, since these reaction products may actually be formed in food products containing unsaturated aldehydes and H_2S or CH_3SH. The reaction mixtures possess strong, and often characteristic, odours.

Reactions products or methanethiol with 2-alkenals and 1-octene-3-one

The reaction products of 2-alkenals with methanethiol have flavour thresholds in the magnitude $\mu g \cdot litre^{-1}$. 3-Methylthiobutanal displays a cheese-like flavour. It is notable that 3-methylthiopropanal ('methional') is also known to possess this flavour. The higher terms have odours varying from unripe tomatoes to bast-like/ floral.

The reaction product of 1-oct en-3-one and methanethiol possesses a typical radish-like flavour.

Reaction products of H_2S and 2-alkenals

The reaction products of 2-alkenals with hydrogen sulphide have flavour thresholds which may be even below $\mu g \cdot litre^{-1}$ in water. Their odours vary from onion-like for the lowest term studied to floral and grapefruit-like for the higher ones.

The flavour impression of the reaction mixture of 2-butenal with hydrogen sulphide depended on the pH during the reaction. In acidic solution, the reaction mixture smelled like onion and leek, but in neutral solution, the omelette odour of 3-mercaptobutanal dominated.

The extract of the reaction products formed from 2-butenal and hydrogen sulphide in dilute aqueous solution at pH 4 was analysed on a glass capillary column coated with SE 30 at 110 °C. The compounds were split in the proportion 1:10 to

Table 3. Results of sensory evaluation of reaction products.

Reagents	Flavour threshold in µg/l in solvent:			Flavour impression
	water	paraffin	milk	
2-butenal+H_2S	0.1	3	30	onion, leek
2-hexenal+H_2S	0.6	2	14	floral (lantana) rhubarb
2-nonenal+H_2S	0.5	6	12	bast-like, floral
2-butenal+CH_3SH	0.5	0.5	5	cheese-like (brie)
2-hexenal+CH_3SH	5	50	75	cabbage, rubbery
2-heptenal+CH_3SH	5	80	40	unripe tomato
2-nonenal+CH_3SH	3	20	30	bast-like, slightly floral
1-octen-3-one+CH_3SH				radish-like

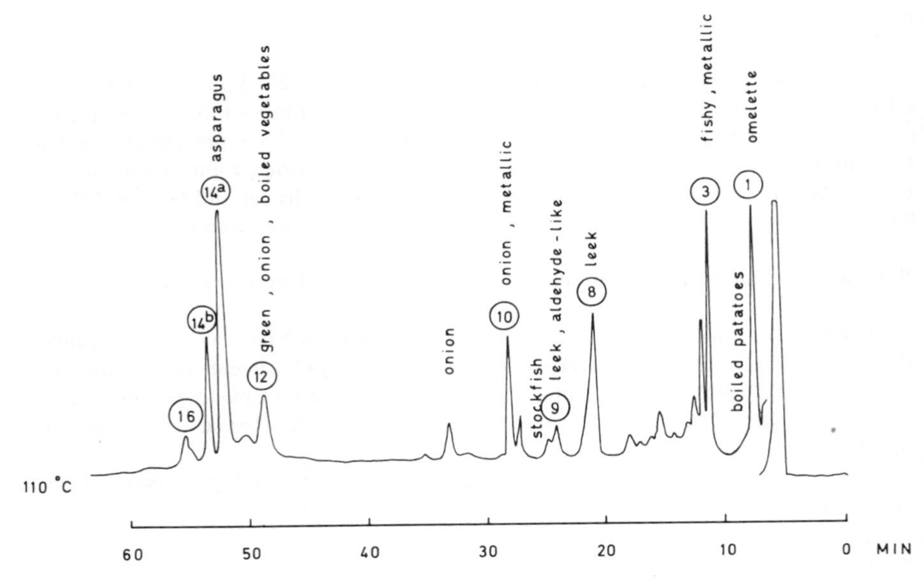

Fig. 4. Gas chromatogram of the 'pH 4' reaction mixture of 2-butenal and H_2S. Peak numbers correspond to those in Table 2; odour assessments are indicated.

the flame ionization detector and a heated glass outlet, respectively, so that a chromatogram and a flavour impression could be obtained simultaneously (Figure 4).

Discussion

Strongly odorous compounds are formed when methanethiol or hydrogen sulphide react with 2-alkenals or with 1-octen-3-one.

Reactions with methanethiol to the formation of the corresponding 3-methylthio carbonyls, and those of hydrogen sulphide with 2-alkenals take place in a more complex manner. This phenomenon was studied extensively for the reaction with 2-butenal.

Many products were obtained, but it was very difficult to prove whether a compound had been formed as a primary reaction product or as a secondary product during analysis. Several of the secondary products were formed during heating, e.g. during gas chromatography.

Since the reactions also occur in dilute aqueous solutions, they can also be expected in many food products that contain both unsaturated 2-alkenals as a result of enzymic oxidation or autoxidation and -SH compounds as a result of heating. The secondary products may also be formed during processing of food products (e.g. sterilization).

We would mention one example. When butter-fat from oxidized butter with a typical fishy flavour was heated for 1 h at 80 °C with a small amount ($0.5\ mg \cdot l^{-1}$) of

H_2S, the fishy off-flavour disappeared completely and changed into a pleasant smell of frying.

A small change in the pH of the reaction mixture in one unit, dramatically changed the composition of the reaction mixture. The composition of the reaction mixture obtained at pH 7 differs particularly from that at pH 6. Therefore, according to the pH of a food product, the presence of 2-butenal and hydrogen sulphide will result in different reaction products.

It was quite interesting to study the flavour quality of the reaction mixtures. The thresholds were low and sometimes the mixtures possessed flavours typical of certain food products.

Acknowledgment

The authors thank J. G. Wassink, J. J. G. van der Pol and H. G. J. Teisman of the Netherlands Institute for Dairy Research and Drs J. Schaefer, Drs R. Belz, C. van Ingen, Ms C. A. Visscher and J. Bouwman of the Central Institute for Nutrition and Food Research TNO for their valuable assistance in performing the infrared and mass spectrometry, and the gas chromatography.

References

Asinger, F. & M. Fischer, 1967. Zur Kenntnis der Dihydrometathiazine-Δ^3. J. prakt. Chem. 35: 81–90.
Badings, H. T., 1967. Causes of ribes flavor in cheese. J. Dairy Sci. 50: 1347–1351.
Badings, H. T., 1975. Wide bore glass capillary columns for G. C., their preparation and modes of application. Chromatographia, in press.
Boelens, M., L. M. van der Linde, P. J. de Valois, H. M. van Dort & H. J. Takken, 1974. Organic sulfur compounds from fatty aldehydes, hydrogen sulfide, thiols and ammonia as flavor constituents. J. agric. Food Chem. 22: 1071–1076.
Firmenich, S. A., 1972. Swiss Patent Appl. 11441/72.
Gershbein, L. L. & C. D. Hurd, 1947. Reaction of hydrogen sulphide with acrylonitrile, acrylic ester and crotonaldehyde. J. Amer. Chem. Soc. 69: 241–242.
Givaudan, S. A., 1972. Swiss Patent Appl. 5353/72.
Kleipool, R. S. C., A. Tas, H. Maarse, R. Neeter & H. T. Badings, Z. LebensmUnters. Forsch., in press.
Maarse, H. & M. C. ten Noever de Brauw, 1972. Kattepiesgeur als luchtverontreiniging. Chem. Weekbl. 68: 11–13.
Maarse, H. & M. C. ten Noever de Brauw, 1974. Another catty odour compound causing air pollution. Chem. Ind.: 36–37.
Moinas, M., 1973. Aroma volatil de produits laitiers; extraction et identification. Mitt. Geb. Lebensmittelunters. Hyg. 64: 60–65.
Patton, S. & D. V. Josephson, 1957. A method for determining significance of volatile flavor compounds in foods. Food Res. 22: 316–318.
Pearce, T. J. P., J. M. Peacock, F. Aylward & D. R. Haisman, 1967. Catty odours in foods. Reactions between hydrogen sulphide and unsaturated ketones. Chem. Ind.: 1562–1563.
Polak's Frutal Works, 1967. U.S. Patent 3, 645, 754.
Schutte, L., 1974. Precursors of sulfur-containing flavor compounds. Crit. Rev. Food Technol. 4: 457–505.
Straten, S. van & F. de Vrijer, 1973. Lists of volatile compounds in food. 3rd ed. Report R no. 4030, Central Institute for Nutrition and Food Research TNO, Zeist, the Netherlands.

Proc. int. Symp. Aroma Research, Zeist, 1975. Pudoc, Wageningen.

Aroma compounds formed by enzymic co-oxidation

W. Grosch, G. Laskawy and K.-H. Fischer

Deutsche Forschungsanstalt für Lebensmittelchemie, 8 München 40, Leopoldstrasse 175, Bundesrepublik Deutschland

Abstract

Volatile compounds which result from the enzymic degradation of unsaturated fatty acids or carotenoids in the presence of oxygen are found in the aromas of many fruits and vegetables. The author suggests that these aroma substances are formed through enzymic co-oxidation reactions. Among the enzymes which occur in plant tissues, the lipoxygenase, H_2O_2-peroxidase and NADP-ferredoxin reductase have so far been published in the literature as generators of peroxyl, hydroxyl or superoxide anion radicals. These radicals can either directly or through disproportionation with the formation of singlet oxygen, start a lipid peroxidation. To check this hypothesis the following model systems were investigated: linolenic acid/alkaline lipoxygenase (soya), – / pH 6.5 lipoxygenase (peas, soya), – /haem catalyst, – / singlet O_2 and β-carotene/linoleic acid/pH 6.5 lipoxygenase (soya).

The volatile compounds (mainly carbonyl compounds), produced during the incubation are reported upon and discussed.

Products of enzymic catalysis

The exclusiveness of enzymic reactions is particularly impressive. Their substrate specificity has been stressed. Enzymes accelerate a definite reaction with a clear stoichiometry. Side reactions do not occur.

In the last few years exceptions to the principle of clear stoichiometry have become known among enzymes that use oxygen as substrate, i.e. *oxidases* and *oxygenases*. Oxygen is a difficult substrate for enzymes. In its ground state it is a poorly reactive bi-radical that is reduced by the oxidases in either two or four 1-electron steps to H_2O_2 or H_2O.

$$O_2 \xrightarrow{1e} \underset{\text{superoxide radical anion}}{O_2^{\ominus}} \xrightarrow{1e;\,2H^{\oplus}} H_2O_2 \xrightarrow{1e;\,H^{\oplus}} H_2O + \cdot OH \xrightarrow{1e;\,H^{\oplus}} 2\,H_2O$$

(Reaction 1)

A 1-electron transfer can also characterize the oxygenases that incorporate one or two atoms of oxygen into one molecule of an organic compound.

Very reactive intermediates arise by 1-electron transfer from oxygen (Reaction 1). The first reduction product, the superoxide radical anion, is not suitable for co-oxidation but can easily disproportionate into H_2O_2 and singlet O_2 (Khan, 1970).

$$2O_2^{\ominus} + 2H^{\oplus} \longrightarrow H_2O_2 + {}^1O_2 \qquad \text{(Reaction 2)}$$

In fact, singlet O_2 is the reactive reagent because it can peroxidize unsaturated compounds such as linoleic and linolenic acids.

In contrast to the superoxide radical anion the third product of oxygen reduction, the hydroxyl radical, is extremely reactive (Hamilton, 1974). In oxygenase catalysis, the number of reactive intermediates can be multiplied by the involvement of the organic substrate. Alkyl, alkoxyl and alkperoxyl radicals can result.

In the catalysis by some oxidases and oxygenases, under particular conditions, such side-reactions can occur. With an excess of oxygen, the enzyme no longer reacts in a specific way. The reactive intermediates of catalysis are formed in such high concentrations that besides the main pathway uncontrolled co-oxidation reactions occur.

This effect of an enzymically initiated co-oxidation seems meaningful in the formation of some significant flavour substances in plant foodstuffs. In fruits and vegetables, various volatile aldehydes and related alcohols and esters arise by enzymic oxidative splitting of unsaturated fatty acids (Drawert et al., 1966, 1973; Kazeniac & Hall, 1970; Grosch & Schwarz, 1971). To explain the formation of the carbonyl compounds, we suggest that they do not arise directly from the enzymic degradation of unsaturated fatty acids but from a lipid co-oxidation. The radicals and singlet oxygen, perhaps initiated by some enzyme activities, would peroxidize linoleic and linolenic acid.

Experimental co-oxidation by lipoxygenases

An enzyme which occurs widely in plant tissues and which is identified as a radical producer (de Groot et al., 1973) is the lipoxygenase (EC 1.13.11.12). Hamberg & Samuelsson (1967) have demonstrated the high specificity of the enzyme. Only compounds like linoleic and linolenic acid which have a pair of methylene group-interrupted double bonds located between ω-6 and ω-10 are oxidized. Under aerobic conditions conjugated *cis,trans*-hydroperoxides (LOOH) are the main products. But the lipoxygenases differ in their optimum pH and their hydroperoxidation specificity (review by Hamberg et al., 1974).

To check our hypothesis about the formation of volatile carbonyl compounds by an enzymic lipid co-oxidation, we first investigated model systems with different lipoxygenases. A fatty acid emulsion was degassed for some minutes with oxygen and then incubated with the enzyme for 20 min at 10 °C. The enzymic reaction was stopped by precipitation of the substrate and of the oxygenated fatty acids with Ca^{2+} or Pb^{2+}. The carbonyl compounds were isolated and identified as 2,4-dinitrophenylhydrazones (2,4-DNP) (Grosch et al., 1974). This analytical procedure is suitable for exact determination of the aldehydes and ketones formed. Against this, the reaction with the 2,4-DNP reagent results in a rearrangement of the *cis*-3 to the *trans*-2-enal (Badings, 1970).

Soya bean contains three lipoxygenase-isoenzymes which can be separated by

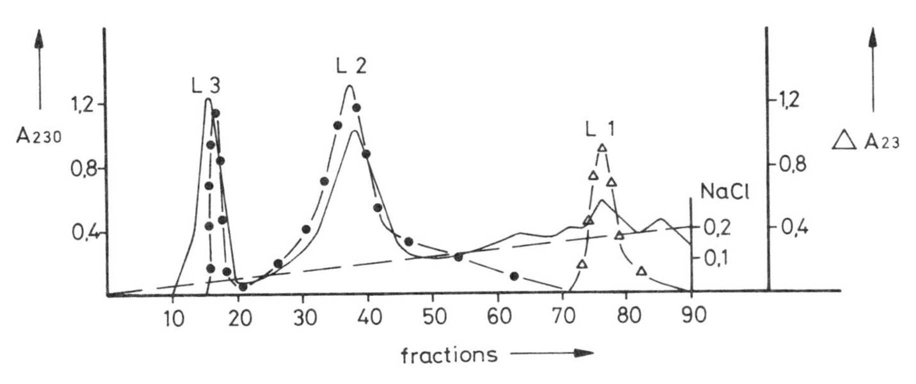

Fig. 1. Extract from soya beans chromatographed on diethylaminoethyl-cellulose. Protein was measured by absorbance at 230 nm. The increase in absorbance at 234 nm between 30 and 60 s was estimated in the test for lipoxygenase activity. Further experimental details are described by Weber et al. (1974).
● — ● — ● Lipoxygenase activity at pH 6.5
△ — △ — △ Lipoxygenase activity at pH 9.0
———— Protein

chromatography on diethylaminoethyl-cellulose (Fig. 1). Lipoxygenases partially isolated by this procedure were incubated with linolenic and linoleic acids at their optimum pH values.

Lipoxygenases from potato and from wheat were also investigated in this way. The results are given in Table 1. Let us first consider the soya enzymes. The oxidation of both of the fatty acids used as substrates was measured.

The enzymes L-2 and L-3 labile at pH 6.5 denatured during reaction for 20 min. Their fatty acid turnover was therefore less than that of the stable L-1. Although L-2 and L-3 oxidized less substrate than L-1, they formed more carbonyl compounds than the 'alkaline' lipoxygenase (Table 1). This difference was particularly evident in the experiment with linolenic acid as substrate. L-1 formed only 1.5 μmol while L-2 and L-3 each formed about 7 μmol of carbonyl compounds.

The literature has assumed that carbonyl compounds form by breakdown of hydroperoxides. Hence the experiment with L-1 should have resulted in the greatest amount of carbonyl compounds because L-1 oxidized considerably more unsaturated fatty acid than L-2 and L-3 (Table 1) and the main products of L-1 catalysis were hydroperoxides (Hamberg & Samuelsson, 1967). However our results do not support this hypothesis since L-2 and L-3 formed more carbonyl compounds than did L-1. Clearly the carbonyl compounds did not arise by breakdown of hydroperoxides but by the reaction sequence of peroxyl radicals generated by the 'neutral' lipoxygenases L-2 and L-3.

Table 1 shows that during the oxidation of linolenic acid the lipoxygenase from potatoes and wheat form carbonyl compounds only to a small extent. But it is not only in this feature that the enzymes from wheat and potatoes behave like the alkaline lipoxygenase L-1. The large hydroperoxidation specificity is also common to all three (Table 1). The wheat and the potato enzymes form predominantly the 9-hydroperoxide and the soya bean enzyme L-1 forms the 13-hydroperoxide. In

Table 1. Amounts of volatile carbonyl compounds formed by lipoxygenases.

Lipoxygenase protein in the reaction system (mg)		Substrate[1]	Conversion of substrate[2] (%)	Monocarbonyl formed (μmol)	Hydroperoxidation specificity (%)			
					C-9	C-12	C-13	C
L-1 soya, pH 8.5	4.2	18 : 2	81	1.5	15[3]		85[3]	
L-2 soya, pH 6.5	3.3		42	3.6	55[3]		45[3]	
L-3 soya, pH 6.5	2.6		56	3.3				
L-1 soya, pH 8.5	5.3	18 : 3	85	1.5	10[3]		90[3]	
L-2 soya, pH 6.5	2.0		32	6.9			6[3]	4
L-3 soya, pH 6.5	2.0		30	7.1	45	5		
K potato, pH 5.5	12.8		56	1.5	98		2	
W wheat, pH 5.8	9.8		43	1.3	78		22	

1. 18 : 2 linoleic acid, 18 : 3 linolenic acid.
2. From diene absorption.
3. Roza & Francke, 1973.

contrast to this the neutral soya bean enzymes appear more to catalyze an autoxidation process, because to a first approximation they form equal amounts of the 9 and 13-hydroperoxides from linoleic acid respectively 9 and 16-hydroperoxides from linolenic acid. The differences from an autoxidation arise in the substrate specificity and the very high reaction velocity of the neutral enzymes.

Identification of the carbonyls

The carbonyl compounds formed from the three soya lipoxygenases were identified. The results in Table 2 show that the 'neutral' enzymes differ from the alkaline lipoxygenase in the number, proportion and structure of volatile products. L-2 and L-3 form a similar range of carbonyl compounds. Propanal predominates with noticeable quantities of *trans*-2,*cis*-4-heptadienal, *trans*-2-hexenal and *trans*-2-pen-

Table 2. Composition of the volatile carbonyl compounds formed by lipoxygenases.

	Fraction in total identified carbonyls (mol %)			
	L-3 (soya)	L-2 (soya)	L-1 (soya)	K (potato)
Acetaldehyde	n.d.	1.5	n.d.	2
Propanal	41	46	18	11
trans-2-Pentenal	11	8	5	3
trans-2-Hexenal	9	9	77	15
trans-2,*cis*-6-Nonadienal	2.5	n.d.	n.d.	46
trans-2,*cis*-4-Heptadienal	20	20.5	n.d.	8
3,5-Octadien-2-one[1]	8	8	n.d.	9
2,4,6-Nonatrienal[1]	8.5	7	n.d.	6

1. Two geometric isomers:
n.d. Not detectable.

tenal. In addition 3,5-octadien-2-one and 2,4,6,-nonatrienal arise in about the same proportions. In contrast L-1 forms mainly *trans*-2-hexenal.

A greater range of volatile aldehydes is formed by the potato enzyme with *trans*-2,*cis*-6-nonadienal as main product.

The results shown relate to semi-purified lipoxygenases. There may have been traces of haem compounds in the enzyme preparation, which could affect the breakdown of peroxides to carbonyl compounds. Because of this the incubation was repeated with a pure lipoxygenase of pH optimum 6.5. This enzyme was isolated from peas by precipitation with $(NH_4)_2SO_4$, gel filtration and ion-exchange chromatography on carboxymethyl-Sephadex and DEAE-cellulose (Arens et al., 1973). Additionally the linolenic acid substrate was peroxidized in a further experiment with haemoglobin as catalyst. The results (Table 3) alongside those of Table 2 indicate that the pure lipoxygenase from peas oxidized the linolenic acid to the same carbonyl compounds as L-2 and L-3.[1]

The structure of the identified volatiles formed with haemoglobin is similar to that of the lipoxygenases. However, there is a big difference in rate of reaction. The system with the haem catalyst had to be incubated for more than 70 h at room temperature for the same amount of carbonyls to be formed as in the experiments with the lipoxygenase from peas. The lipoxygenase catalysis differs from the haem catalysis also in the main volatile carbonyls formed (Table 3).

In conclusion, these experiments showed that neutral lipoxygenases influence the formation of aroma-active carbonyl compounds from unsaturated fatty acids. In contrast to the alkaline enzymes, they allow co-oxidation of a lager proportion of peroxyl radicals first formed. Removal of H atoms from other substrate molecules and their attachment near the active site of the enzyme, formation of oxyl radicals by the combination of two peroxyl radicals, and other reaction sequences result in a variety of carbonyl compounds.

Table 3. Oxidation of linolenic acid by a purified lipoxygenase from peas (I) and haemoglobin (II).

Identified carbonyls	Fraction in total identified carbonyls (mol %, (mol / mol) x 100)	
	I	II
Acetaldehyde	2.5	5
Propanal	59	39
trans-2-Butenal	trace	trace
trans-2-Pentenal	10	6.5
trans-2-Hexenal	2.5	20.5
trans-2,*cis*-4-Heptadienal	20	6
trans-2,*cis*-6-Nonadienal		2.5
3,5-Octadien-2-one	6	20.5

1. No statement can be made about the formation of the 2,4,6-nonatrienals with the pea lipoxygenase and haem catalysis. Only the compounds shown in Table 3 were identified.

Alternative carbonyl formation

During catalysis by lipoxygenases of pH optimum 6.5 only traces of volatile aldehydes form, the main products being a mixture of oxidized fatty acids (Arens & Grosch, 1974). Because of these low concentrations, large catalytic concentrations are necessary for the development of sensorily detectable concentrations of the aldehydes. In some Leguminosae, potatoes, egg plant and artichoke, such high catalytic concentrations occur, but in foodstuffs such as apples, tomatoes and cucumbers much less is present (Pinsky et al., 1971). However in these fruit and

Table 4. Products of the reaction of linoleic acid with singlet-oxygen.

Identified carbonyls	Fraction in total identified carbonyls (mol %, (mol / mol) x 100)
Propanal	46.5
trans-2-Pentenal	trace
trans-2-Hexenal	42
trans-2,*cis*-4-Heptadienal	2
trans-2,*cis*-6-Nonadienal	10

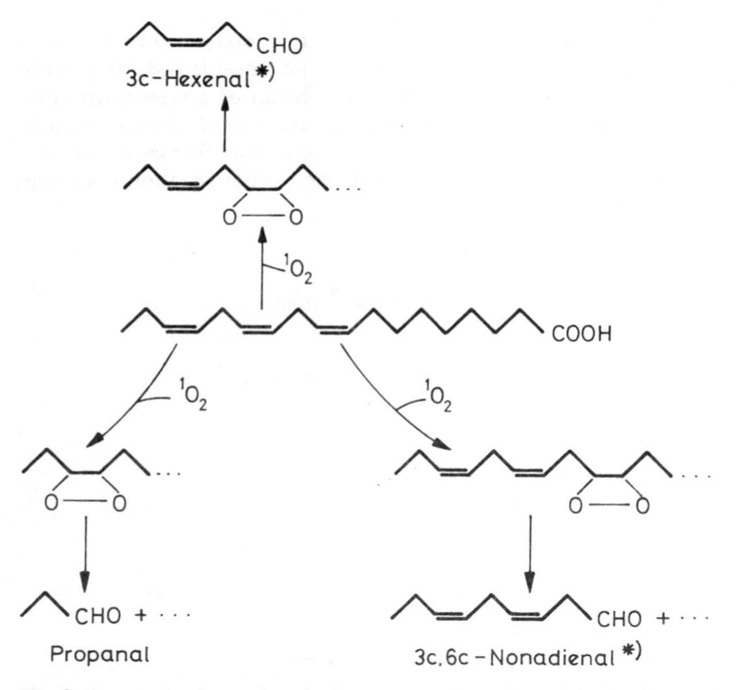

Fig. 2. Proposed scheme for the formation of volatile aldehydes during the oxidation of linolenic acid with singlet oxygen. * The *cis*-3 is isomerized to the *trans*-2-aldehyde through derivative formation with 2,4-DNPH.

vegetables, sensorily detectable concentration of volatile aldehydes also arise from unsaturated fatty acids. We suggest that in such cases further enzymic co-oxidation leads to formation of carbonyl compounds. Perhaps singlet O_2 arises in these plant tissues through the pathway, Reactions 1 and 2, given in the introduction. To elucidate whether aroma substances could be so formed, we investigated which volatile carbonyls were formed from unsaturated fatty acids in a model experiment with singlet O_2 generated by hydrolysis of K_3CrO_8. Only three aldehydes arose in detectable concentrations from linolenic acid (Table 4). They are essential components of the aroma of cucumber (Forss et al., 1962).

To explain the formation of the volatile aldehydes we suppose that besides an 'ene' mechanism (Foole, 1968) the reaction shown in Figure 2 occurs. Each double bond of the linolenic acid can add one molecule of singlet oxygen. It results in dioxethanes which then decompose. As well as other compounds, the three volatile aldehydes form.

Oxidation of β-carotene

The lipoxygenases differ not only in optimum pH values and in peroxidation specificity but also in carotene bleaching activity (Weber et al., 1974).

The neutral lipoxygenases from peas and soya beans can co-oxidise with a high velocity β-carotene in the presence of linoleic acid and oxygen.

In contrast, the alkaline enzyme L-1 from soya beans possesses a considerably lower bleaching activity. The co-oxidation of carotenoids leads not only to bleach-

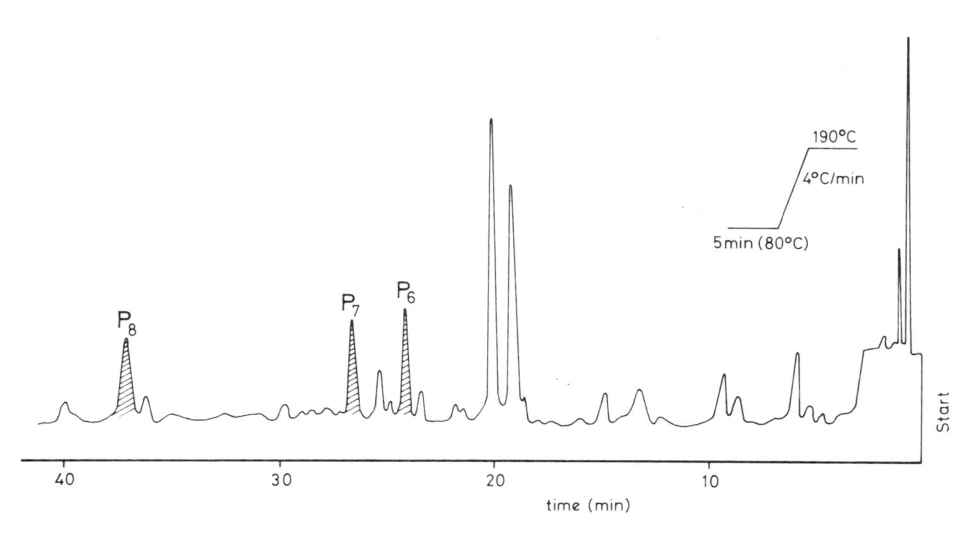

Fig. 3. Gas chromatography of volatiles from the model system β-carotene, linoleic acid, lipoxygenase from soya beans. Column: 3 m 3 % SILAR 5 CP on Gas Chrom Q (100 – 120 mesh). Temperature programme.
Carrier gas: Helium. Flow rate: 25 ml/min. P_6, P_7 and P_8 are volatiles which are formed by the co-oxidation of β-carotene.

ing but also to the formation of aroma substances. We have investigated the model system β-carotene/linoleic acid/lipoxygenase (soya beans). After an incubation of 3 minutes at 15 °C the reaction products were extracted and gas chromatographically analysed. Besides from the volatile compounds that arise from linoleic acid three compounds P_6, P_7 and P_8 appear (Fig. 3). They are only visible in experiments with β-carotene.

P_6 was identified as β-ionone. The analysis of P_7 and P_8, which are also volatile carbonyl compounds, is still being worked on.

The compounds P_7 and P_8 have no strong flavour. The most important aroma substance that arise from bleaching of the β-carotene is the β-ionone.

Conclusion

The model experiments show that the hypothesis formulated in the introduction is supported by the experiments. However, many questions remain open. Which enzymes can start a co-oxidation? How great is the role of the different lipoxygenases in the aroma formation of a distinct fruit or vegetable in an oxidative breakdown of unsaturated fatty acids? And in how far do other enzymes take part in this process?

References

Arens, D., W. Seilmeier, F. Weber, G. Kloos & W. Grosch, 1973. Purification and properties of a carotene co-oxidizing lipoxygenase from peas. Biochim. biophys. Acta 327: 295–305.

Arens, D. & W. Grosch, 1974. Non-volatile reaction products from linoleic acid. Comparison of a ground pea suspension with a purified pea lipoxygenase. Z. LebensmUnters. Forsch. 156: 292–299.

Badings, H. T., 1970. Cold-storage defects in butter and their relation to the autoxidation of unsaturated fatty acids. Neth. Milk Dairy J. 24: 147–256.

de Groot, J. J. M. C., G. J. Garssen, J. F. G. Vliegenthart & J. Boldingh, 1973. Detection of linoleic acid radicals in the anaerobic reaction of lipoxygenase. Biochim. biophys. Acta 326: 279–294.

Drawert, F., W. Heimann, R. Emberger & R. Tressl, 1966. Enzymatische Bildung von Hexen-2-al-(1), Hexanal und deren Vorstufen. Liebig's Ann. Chem. 694: 200–208.

Drawert, F., R. Tressl, W. Heimann, R. Emberger & M. Speck, 1973. Ueber die Biogenese von Aromastoffen bei Pflanzen und Früchten 15. Enzymatisch-oxydative Bildung von C_6-Aldehyden und Alkoholen und deren Vorstufen bei Äpfeln und Trauben. Chem. Mikrobiol. Technol. Lebensm. 2: 10–22.

Foote, C. S., 1968. Photosensitized oxygenations and the role of singlet oxygen. Accounts Chem. Res. 1: 104–110.

Forss, D. A., E. A. Dunstone, E. H. Ramshaw & W. Stark, 1962. The flavor of cucumbers. J. Food Sci. 27: 90–93.

Grosch, W. & J. M. Schwarz, 1971. Linoleic and linolenic acid as precursor of the cucumber flavor. Lipids 6: 351–352.

Grosch, W., G. Laskawy & K.-H. Fischer, 1974. Oxidation of linolenic acid in the presence of haemoglobin, lipoxygenase or by singlet oxygen. Identification of the volatile carbonyl compounds. LebensmittWiss. Technol. 7: 335–338.

Hamberg, M. & B.,Samuelsson, 1967. On the specificity of the oxygenation of unsaturated fatty acids catalyzed by soy bean lipoxygenase. J. biol. Chem. 242: 5329–5335.

Hamberg, M., B. Samuelsson, I. Björkhens & H. Danielsson, 1974. Oxygenases in the fatty acid and steroid metabolism. In: O. Hayaishi (ed.), Molecular mechanism of oxygen activation. Academic Press.

Hamilton, G. A., 1974. Chemical models and mechanisms for oxygenases. In: O. Hayaishi (ed.), Molecular mechanisms of oxygen activiation. Academic Press.
Kazeniac, S. J. & R. M. Hall, 1970. Flavor chemistry of tomato volatiles. J. Food Sci. 35: 519–530.
Khan, A. U., 1970. Singlet molecular oxygen from superoxide anion and sensitized fluorescence of organic molecules. Science 168: 476–477.
Pinsky, A., S. Grossman & M. Trop, 1971. Lipoxygenase content and antioxidant activity of some fruits and vegetables. J. Food Sci. 36: 571–572.
Roza, M. & A. Francke, 1973. Product specificity of soy bean lipoxygenases. Biochim. biophys. Acta 316: 76–82.
Weber, F., G. Laskawy & W. Grosch, 1974. Co-oxidation of carotene and crocin by soybean lipoxygenase isoenzymes. Z. LebensmUnters. Forsch. 155: 142–150.

Proc. int. Symp. Aroma Research, Zeist, 1975. Pudoc, Wageningen.

Analysis of off-flavours in food

J. M. H. Bemelmans and M. C. ten Noever de Brauw

Central Institute for Nutrition and Food Research TNO, Utrechtseweg 48, Zeist, the Netherlands

This paper describes off-flavours in food and the main factors causing their formation during production, processing and handling of food products. Subsequently, the objectives of analysis of off-flavours are dealt with and its various aspects are compared with those of normal flavour analysis. Actually the investigation of off-flavours constitutes a 'character impact' study.

Modern mass spectrometry is most important to the identification of off-flavour compounds. An example is given of analysis of a chicken feed suspect of causing a musty taint in eggs. The feed was found to contain chloroanisoles.

Introduction

Webster's Third New International Dictionary defines an off-flavour as "a flavour that is not natural or up to standard owing to deterioration or contamination". Off-flavours, often indicated as taints, are mostly caused by substances that are not naturally present in a product. These substances may be formed from food constituents, but usually they enter the food by contamination. It is also possible that off-flavours result from serious disturbance of the balance of the various substances constituting the flavour (Arthey, 1974). If so, the concentration of substances naturally present in a product is either too high or too low, which affects the subtlety of the flavour. This usually results in a poor organoleptic quality though seldom in a serious off-flavour.

Off-flavours sometimes render a food inedible, but generally they only reduce its organoleptic and consequently its economic value.

Origin of off-flavours

Some factors liable to induce off-flavours in food are summarized in Table 1. The *production* of raw materials for food can already initiate off-flavours in the final product. The use of pesticides, fungicides, etc., as well as the presence of pollutants in the atmosphere at the site of cultivation may actually lead to off-flavours in plants and fruits (Tanner 1972, 1973). The sex of animals (Patterson, 1968a; Fuchs, 1971) and feed composition may have a strong influence on the flavour quality of products of animal origin. The fishy taint in broilers raised on a diet containing fish-meal is an example (Atkinson et al., 1972; Wessels et al., 1973).

Table 1. Factors which may cause off flavours in food.

Food production	*Food processing*	*Food handling*
Vegetable products:	microbial spoilage	contamination (shipping + storage)
pesticides	heat exposure	oxidation
pollutants	contamination (disinfectants)	
Animal products:	packaging	
sex		
feed		

Likewise the presence of contaminants in feed may endanger the organoleptic quality of animal products. Minute quantities of chloroanisoles in the feed were shown to cause a musty taint in eggs and broilers (Bemelmans & ten Noever de Brauw, 1974).

During the *processing* of raw materials into food products, off-flavours are occasionally formed. This can be caused by microbial spoilage, which can occur during any stage of processing or handling, or by overexposure to heat (Bates, 1970). Heat results in certain chemical reactions and the formation of undesirable compounds (e.g. burnt odour).

Many off-flavours are caused by contamination, despite the food processor's awareness of these dangers. Especially chlorine-containing detergents and disinfectants prove dangerous, since chlorine may react with food substances (e.g. phenols) and form potent odorants (Burttschell et al., 1959).

Packaging materials must be free from substances that could cause taints, since they are in close contact with foods for a considerable length of time. A well known example is the so-called 'catty odour', which is caused by addition of hydrogen sulphide from food to mesityl oxide, which is sometimes present in paints and lacquers. The product, 4-mercapto-4-methylpentan-2-one, is a potent odorant with very low threshold (Badings, 1967; Patterson, 1968b).

Food can become contaminated with substances causing off-flavours during *handling,* especially while being stored and shipped. They may, for instance, come into contact with volatiles from paints, lacquers, chemicals or exhaust gases (Saxby, 1973).

Oxidation of food components may lead to off-flavours, mostly during storage. The best known example is the oxidation of unsaturated fatty acids. The oxidation products, lower aldehydes and ketones, have very low thresholds, and cause rancid flavours (Grosch, 1975; Downey, 1969).

Objectives of off-flavour analysis

It is obvious that the objectives of off-flavour analysis vary according to the type, frequency and origin of the off-flavour (Table 2). Identification of substances causing the off-flavour reveals their physical properties, whence one can ascertain whether the substances can be removed from the food to upgrade its quality. If removal is impossible or too expensive, it is important to know whether the substances responsible are toxic. Often it is even more important to determine how the off-flavour is formed. This knowledge is indispensable because it allows precautions against future occurrence. Information from such analysis is often useful too in settling disputes over financial damages resulting from an off-flavour.

Off-flavour and flavour research

First let us compare the various aspects of the analysis of off-flavours with those of common flavour analysis, before we consider procedures used for taint studies (Table 3). In both types of analysis, the substances of interest are usually present at low concentrations. The number of compounds that contributes to food flavours is large, though still limited. Some may be very important to a flavour, but the investigator has to consider many others too, since they also contribute to the flavour. The number of compounds that can produce an off-flavour is virtually unlimited. A specific off-flavour, however, is usually caused by only one or a few substances, thus the analysis constitutes a 'character impact' study.

Artefact formation, which interferes in the analysis of a flavour, is permitted during off-flavour analysis, as long as the presence of the off-flavour compounds can be traced organoleptically. Organoleptic evaluation is of great importance in both types of study. Flavour scientists hope to ascertain the actual composition of flavours and to obtain knowledge needed to enhance flavours and to create new ones. An objective of off-flavour studies is prevention of future taints.

Table 2. Objectives of off-flavour analysis.

- Removal off-flavour substances
- Toxicity off-flavour substances
- Prevention of future off-flavours
- Liability for damages

Table 3. Off-flavour versus flavour research.

Flavour research:	*Off-flavour research:*
– compounds present in low concentration	– compounds present in low concentration
– large, but limited number of compounds, many of which are usually important	– 'unlimited' number of compounds, only one or a few of which are usually important
– artefact formation should be prevented	– artefact formation not necessarily to be prevented
– organoleptic evaluation necessary	– organoleptic evaluation necessary
– application	– prevention

Analysis of off-flavours

Off-flavour analysis is, by nature, a character impact study. Thus the aim of isolation and concentration procedures is not to obtain a good odour concentrate of the product as a whole, but to obtain a concentrate of components responsible for the off-flavour.

Selective procedures for isolation and concentration can only be used if information is available about the likely identity of the off-flavour compounds. This information is obtained either from the 'case history' of the sample or from the off-flavour itself, if characteristic. With modern mass spectrometry one can check for the suspect components without fractionation or purification of concentrates. The concentrates with the off-flavour are introduced into a high-resolution mass spectrometer and checked by the fixed-mass technique for the presence of characteristic masses.

The off-flavour concentrates can also be examined by a combination of gas chromatograph, mass spectrograph and computer. Mass fragmentography is used to determine the presence of characteristic masses. A mass spectrum and retention time can be obtained simultaneously. If a membrane separator is used, the odour of the suspected compound can also be evaluated by sniffing at the outlet of the interface.

The analysis is more difficult when no information at all is available about the possible identity of the off-flavour substances. During the whole analytical procedure, fractions must be organoleptically tested to determine which are important for the off-flavour. Usually the off-flavour concentrate is fractionated on a packed column and the effluent is evaluated organoleptically. Fractions of interest are either fractionated further on another packed column of different polarity, or transferred directly to a capillary column, coupled to a mass spectrometer. By sniffing at the interface, one must determine what interval of retention time of the chromatogram is important (Maarse & ten Noever de Brauw, 1972). Subsequently, the computer is instructed to produce the mass spectra of this interval, as well as selective mass plots. Access to a computer considerably simplifies identification, as will be illustrated.

One of the problems of off-flavour analysis is that it becomes increasingly difficult to trace organoleptically the off-flavour as analysis proceeds. This is especially true if the off-flavour is faint or if it is caused by interaction of several components, in which case organoleptic confirmation of their presence becomes impossible as soon as they are separated into different fractions.

Part of this problem may soon be solved by the development of new computer programmes, to compare qualitatively the chromatograms of a product of good with one of bad quality. Both chromatograms have to be run under identical conditions. Spectra of compounds present at the same place in the chromatograms are either compared with each other or subtracted from each other. This enables one to trace analytically the difference between the two chromatograms and consequently the presence of extraneous components in a product. However, this procedure does not eliminate the need for organoleptic analysis. One should always assess whether the foreign components identified are responsible for the off-flavour.

Finally, let us describe part of an off-flavour study recently made at this Institute.

In recent years, a musty taint occurred regularly in eggs and broilers. Engel et al. (1966) showed the cause of this taint to be 2,3,4,6-tetrachloroanisole. Curtis et al. (1972, 1974) found that this substance was formed from chlorophenol by fungi present in the litter. Chlorophenols are used as wood preservatives and are often present in wood shavings used in poultry houses (Parr et al., 1974). However, musty eggs were recently obtained from poultry houses where no shavings were used. Again, the mustiness was caused by chloroanisoles (Bemelmans & ten Noever de Brauw, 1974). To determine the source of contamination, feed samples were analysed. They were submitted to a combined procedure of steam distillation and extraction (Nickerson & Likens, 1966). The extracts were concentrated and introduced directly into a double-focusing mass spectrometer (Varian-MAT, type 731). By high-resolution mass spectrometry (resolution 10 000), all samples were tested for the molecular ions of the chloroanisoles, by the fixed-mass technique. The suspect samples proved to contain the parent ions of both trichloro and tetrachloro anisoles. There was no evidence for the presence of pentachloroanisole.

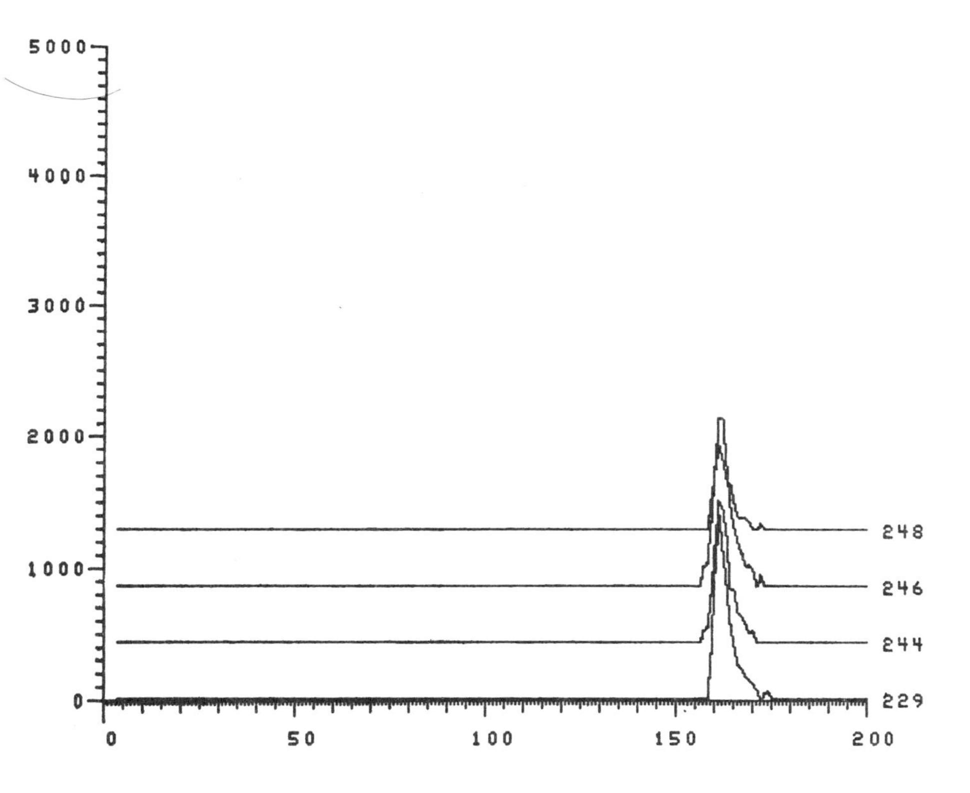

Fig. 1. Selected mass plot of a concentrate of chicken feed of suspect quality. The masses m/e 248, 246, 244 and 229 are characteristic for tetrachloroanisole.

Subsequently, the concentrates were analysed by a combination of gas chromatography mass spectrometry and computer (Varian-MAT) to determine which isomers were present. The concentrates were separated on an Apiezon L capillary column (50 m, wide bore, stainless steel; oven temp. 140 °C). The effluent was

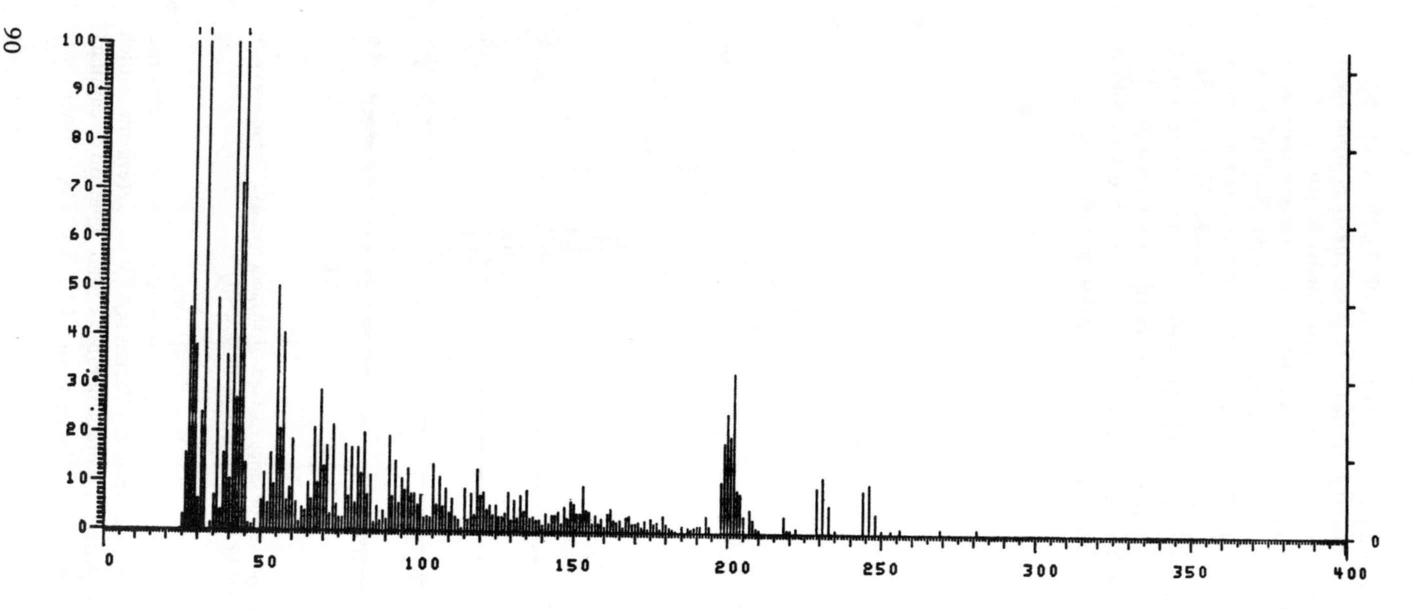

Fig. 2. Uncorrected mass spectrum recorded at Position 163 in the chromatogram.

scanned continuously. Spectra and total ion currents were recorded and stored by the computer. After completion of the analysis, mass fragmentography was used to confirm the presence of trichloro and tetrachloro anisole (Fig. 1).

At Position 163 in the chromatogram, a compound was present with four masses characteristic for tetrachloroanisole. Subsequently, the computer was instructed to plot the uncorrected spectrum recorded there (Fig. 2).

However, interpretation of such a spectrum is difficult, because of the presence of interfering masses. After correction for column bleeding and for interfering components, a spectrum was obtained identical with the reference spectrum of 2,3,4,6-tetrachloroanisole (Fig. 3).
The component under investigation also had the same retention time as 2,3,4,6-tetrachloroanisole and hence was identified positively.

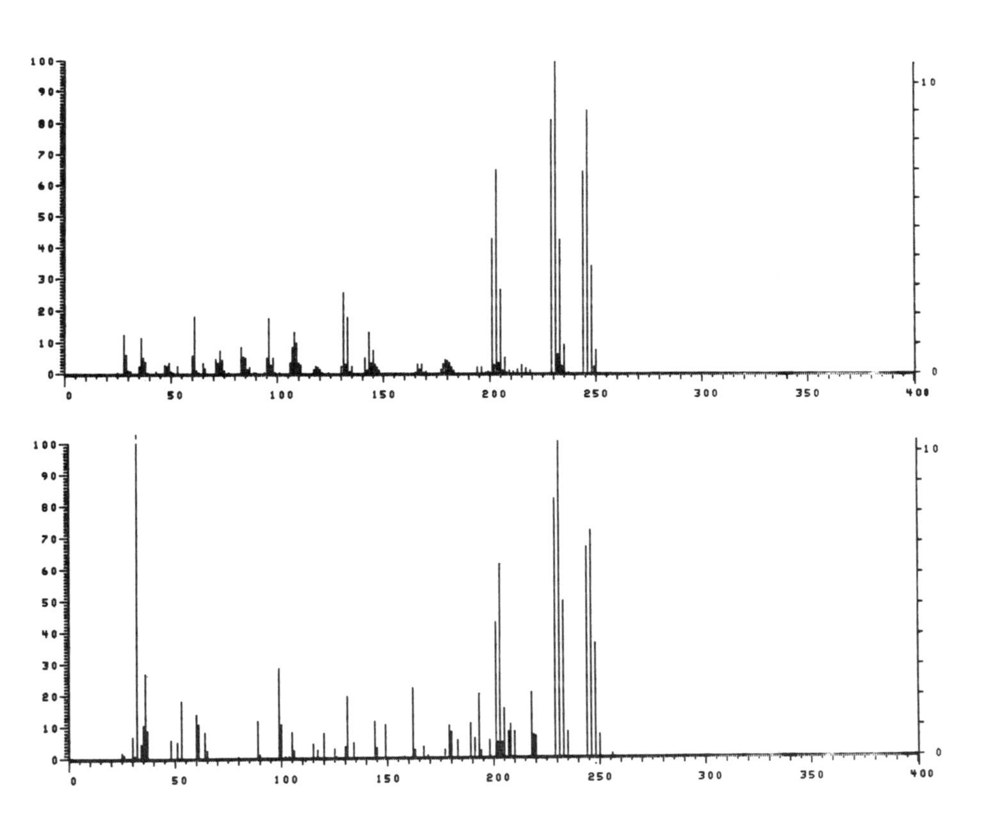

Fig. 3. Mass spectrum at Position 163 in the chromatogram. Top: after correction for column bleeding and interfering compounds. Bottom: the reference spectrum for 2,3,4,6-tetrachloroanisole recorded in the same way.

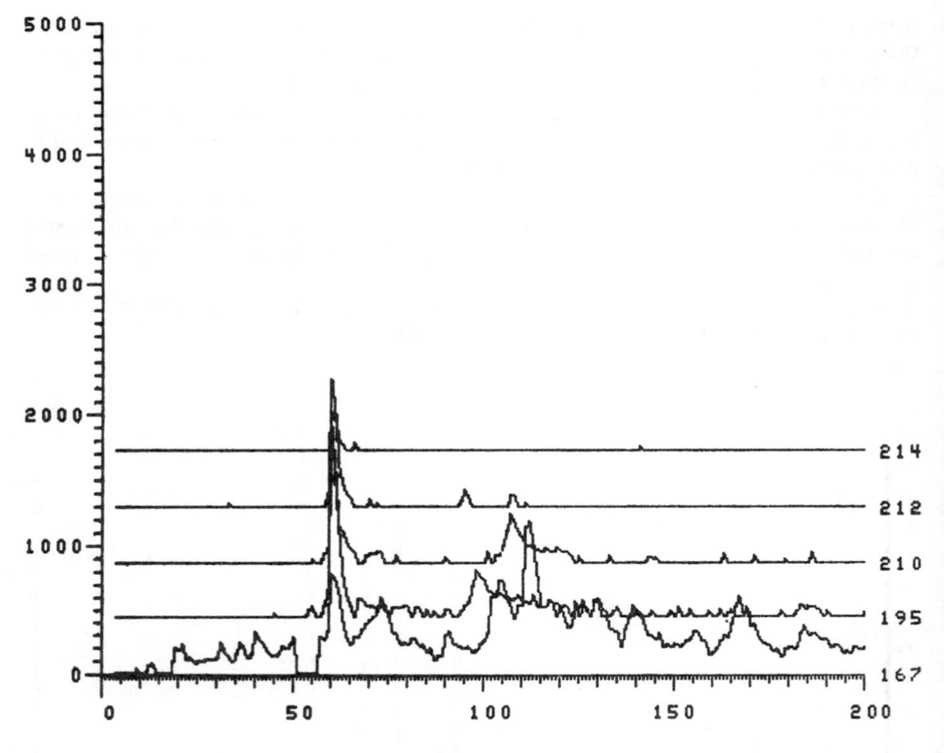

Fig. 4. Selected mass plot for the masses *m/e* 214, 212, 210, 195 and 167, which are characteristic for trichloroanisole.

Also, five masses characteristic for trichloroanisoles were plotted (Fig. 4).

The component at Position 61 was identified as 2,4,6-trichloroanisole in the same manner. The feed of suspected quality proved to be contaminated with chloroanisoles that were musty and had extremely low odour thresholds (Griffiths, 1974).

The mass fractions of anisole were estimated by gas chromatography. Between 0.01 and 0.17 mg/kg of 2,3,4,6-tetrachloroanisole was present in the feed and less than 0.03 mg/kg of 2,4,6-trichloroanisole. Feed of unsuspect quality was found not to contain any chloroanisoles.

Acknowledgment

The authors are grateful to Mr N. L. A. te Loo for the preparation of samples for the analysis and to Mr C. van Ingen and Mr J. Bouwman for assistance in recording and interpreting the mass spectra. Also thanks to Drs R. Belz for checking this manuscript.

References

Arthey, V. D., 1974. Problems associated with the use of pesticides: taints. Pestic. Sci. 5: 113–115.

Atkinson, A., L. G. Swart, R. P. van der Merwe & J. P. H. Wessels, 1972. Flavour studies with different levels and times of fish meal feeding and some flavour-imparting additives in broiler diets. Agroanimalia 4: 53–62.

Badings, H. T., 1967.Causes of ribes flavour in cheese. J. Dairy Sci. 50: 1347–1351.

Bates, R. P., 1970. Heat-induced off-flavour in avocado flesh. J. Food Sci. 35: 478–482.

Bemelmans, J. M. H. & M. C. ten Noever de Brauw, 1974. Chloroanisoles as off-flavour components in eggs and broilers. J. agric. Food Chem. 22: 1137–1138.

Burttschell, R. H., A. A. Rosen, F. M. Middleton & M. B. Ettinger, 1959. Chlorine derivatives of phenol causing taste and odor. J. Amer. Water Works Assoc. 51: 205–214.

Curtis, R. F., D. G. Land, N. M. Griffiths, M. Gee, D. Robinson, J. L. Peel, C. Dennis & J. M. Gee, 1972. 2,3,4,6-Tetrachloroanisole association with musty taint in chickens and microbiological formation. Nature 235: 223–224.

Curtis, R. F., C. Dennis, J. M. Gee, M. G. Gee, N. M. Griffiths, D. G. Land, J. L. Peel & D. Robinson, 1974. Chloroanisoles as a cause of musty taint in chickens and their microbiological formation from chlorophenols in broiler house litters. J. Sci. Food Agric. 25: 811–828.

Downey, W. K., 1969. Lipid oxidation as a source of off-flavour development during the storage of dairy products. J. Soc. Dairy Technol. 22: 154–162.

Engel, C., A. P. de Groot & C. Weurman, 1966. Tetrachloroanisole: a source of musty taste in eggs and broilers. Science 154: 270–271.

Fuchs, G., 1971. The correlation between the 5α- androst-16-ene-3-one content and the sex odour intensity in boar fat. Swed. J. agric. Res. 1: 233–237.

Griffiths, N. M., 1974. Sensory properties of the chloroanisoles. Chem. Senses Flavor 1: 187–195.

Grosch, W., 1975. Ablauf und Analytik des oxydativen Fettverderbs. Z. LebensmUnters. Forsch. 157: 70–83.

Maarse, H. & M. C. ten Noever de Brauw, 1972. Kattepiesgeur als luchtverontreiniging. Chem. Weekbl. 68: 36–37.

Nickerson, G. B. & S. T. Likens, 1966. Gas chromatographic evidence for the occurrence of hop oil components in beer. J. Chromatogr. 21: 1–5.

Parr, L. J., M. G. Gee, D. G. Land, D. Robinson & R. F. Curtis, 1974. Chlorophenols from wood preservatives in broiler house litter. J. Sci. Food Agric. 25: 835–841.

Patterson, R. L. S., 1968a. 5α-Androst-16-ene-3-one: compound responsible for taint in boar fat. J. Sci. Food Agric. 19: 31–37.

Patterson, R. L. S., 1968b. Catty odours in food: their production in meat stores from mesityl oxide in paint solvents. Chem. Ind. 548–549.

Saxby, M. J., 1973. Taints in foodstuffs: Detection and identification. Food Manuf. F. Rev. 19–20 + 24 + 26.

Tanner, H., 1972. Ein Fall von Weinkontamination durch industriebedingte Emissionsprodukte. Mitt. Geb. Lebensmittelunters. Hyg. 63: 60–71.

Tanner, H., 1973. Der Einfluss von atmosphärischen Umweltstoffen auf den Wein. Schweiz. Z. Obst Weinbau 109: 585–591.

Wessels, J. P. H., A. Atkinson, R. P. van der Merwe & J. H. de Jongh, 1973. Flavour studies with fish meals and with fish oil fractions in broiler diets. J. Sci. Food Agric. 24: 451–461.

Proc. int. Symp. Aroma Research, Zeist, 1975. Pudoc, Wageningen.

Organic sulphur compounds as flavour constituents: reaction products of carbonyl compounds, hydrogen sulphide and ammonia

Short communication

H. Boelens, L. M. van der Linde, P. J. de Valois, J. M. van Dort and H. J. Takken

Naarden International N.V., Research Department, P.O. Box 2, Naarden, the Netherlands

Abstract

Carbonyl compounds, hydrogen sulphide, and ammonia are often constituents of food flavours. We therefore studied the reaction products of these compounds by combination of gas chromatography and mass spectrometry. The primary and secondary reactions of the degradation products of cysteine are discussed. Diketones react with ethanal, hydrogen sulphide and ammonia to form thiazolines, thiazoles, oxazoles, imidazoles, trithiolanes and dithianes. With hydrogen sulphide, diketones produce oxothioketones, mercaptoenones, mercaptoketones, ketonic disulphides and 2,5-dialkylthiophenes. Homocyclic and oxygen-containing heterocyclic diketones react with hydrogen sulphide to form mercaptoketones and mercaptans.

Introduction

Nowadays the importance of organic sulphur compounds in food flavours is generally recognized. For references, see Badings & Maarse (1975) and Boelens et al. (1974).

Because carbonyl compounds, hydrogen sulphide and ammonia have been found in many flavours, especially in the aromas of meat, coffee, *Allium* species, rice and animal fats, we studied the reactions between these compounds. The reaction products were analysed by combined gas chromatography and mass spectrometry.

Methods and materials

Reaction of carbonyl compounds with hydrogen sulphide and ammonia were carried out in a 100-cm^3 Carius tube (Pyrex glass, wall thickness 4 mm, maximum pressure 2 MPa), with a Teflon-coated screw lock. During reaction, the tube was protected by a metal sleeve. Mass spectra were recorded on a Varian MAT/CH 5 mass spectrometer coupled to a Varian Aerograph Model 1220 gas chromatograph. Separations were in an all-glass system with capillary columns as previously described (Boelens et al., 1971).

Results and discussion

Primary and secondary reactions of the degradation products of cysteine

The formation of ethanal, hydrogen sulphide, and ammonia from cysteine was studied by Obata & Tanaka (1965) and by Fujimaki et al. (1969). In Figure 1 is a scheme depicted for the possible pathway of degradation of cysteine.

When ethanal was allowed to react with hydrogen sulphide (substance quotients 2:3) in a closed glass vessel, the main products were 1,1-ethanedithiol and bis-(1-mercaptoethyl) sulphide. The latter compound proved to be a key intermediate in the formation of many important flavour components.

bis-(1-Mercaptoethyl) sulphide was easily oxidized to 3,5-dimethyl-1,2,4-trithiolane, which has been found in the aroma of boiled meat (Chang et al., 1968 and Wilson et al., 1973), in beef broth (Brinkman et al., 1972), and in potatoes (Buttery et al., 1970). When bis-(1-mercaptoethyl) sulphide was treated with acids, it disproportionated to 2,4,6-trimethyl-1,3,5-trithiane, which occurs in pressure cooked meat (Wilson et al., 1973).

By heating in the presence of oxygen the bis-(1-mercaptoethyl) sulphide was converted into a mixture of diethyl-disulphide and trisulphide, which also occur in the flavour complex of boiled meat (Wilson et al., 1973).

bis-(1-Mercaptoethyl) sulphide reacted with ammonia to give 2,4,6-trimethyldihydro-1,3,5-dithiazine. 2,4,6-Trimethyldihydro-1,3,5-dithiazine (thialdine) has been found in beef broth (Brinkman et al., 1972) and in boiled meat (Wilson et al., 1973).

Figure 2 is a survey of organic sulphur compounds formed from ethanal with hydrogen sulphide and ammonia. Cysteine can be considered as a possible precursor for all these aroma components.

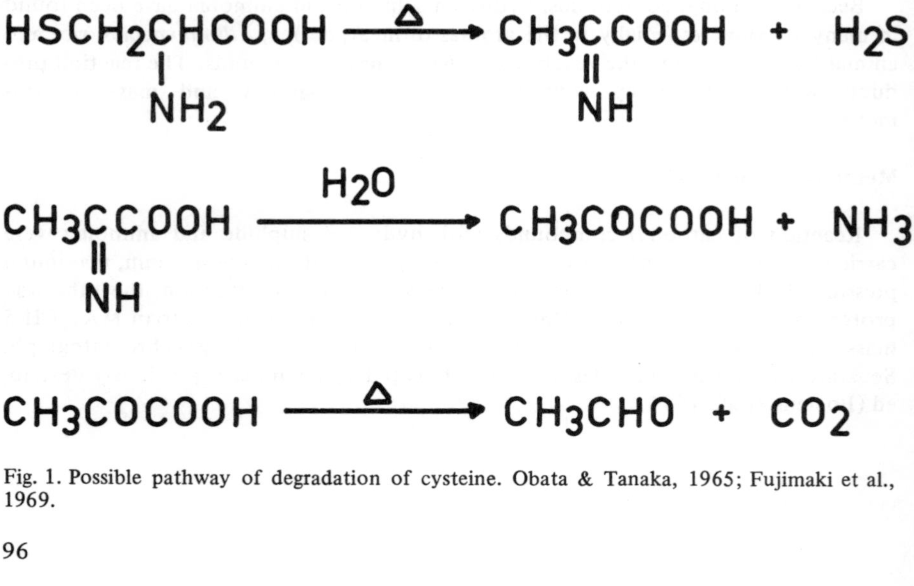

$$HSCH_2CH(NH_2)COOH \xrightarrow{\Delta} CH_3C(=NH)COOH + H_2S$$

$$CH_3C(=NH)COOH \xrightarrow{H_2O} CH_3COCOOH + NH_3$$

$$CH_3COCOOH \xrightarrow{\Delta} CH_3CHO + CO_2$$

Fig. 1. Possible pathway of degradation of cysteine. Obata & Tanaka, 1965; Fujimaki et al., 1969.

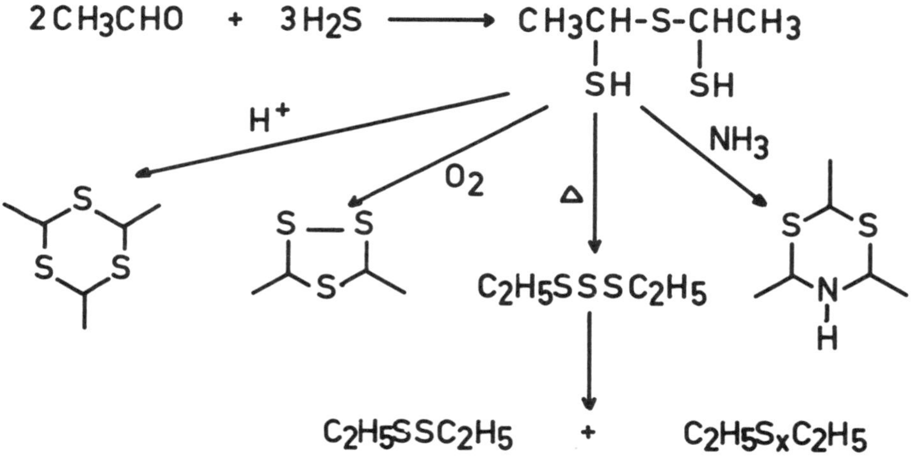

$$2\,CH_3CHO + 3\,H_2S \longrightarrow CH_3CH(SH){-}S{-}CH(SH)CH_3$$

H^+ O_2 Δ NH_3

$C_2H_5SSSC_2H_5$

$C_2H_5SSC_2H_5 + C_2H_5S_xC_2H_5$

Fig. 2. Secondary reaction products of cysteine. Boelens et al., 1974; Ledl, 1975.

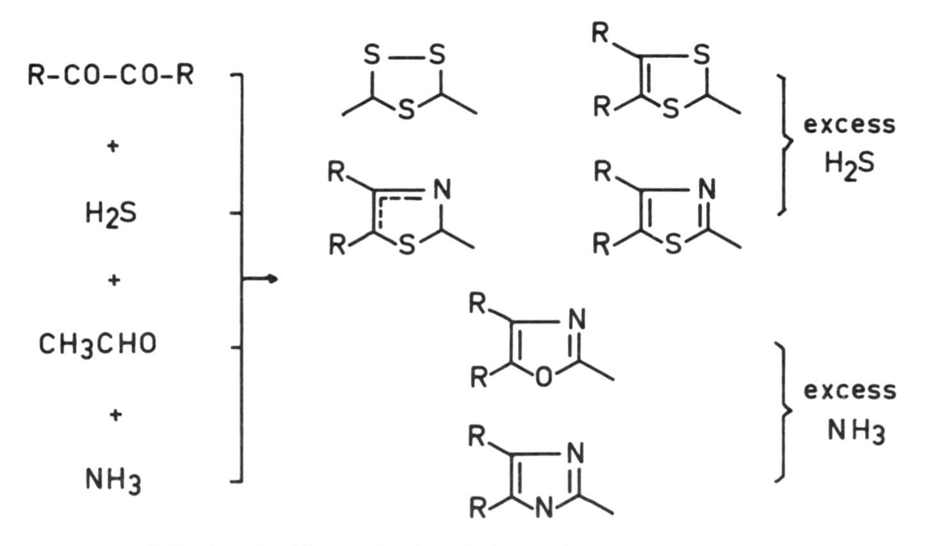

Fig. 3. Vicinal dicarbonyls with cysteine degradation products.

Reaction products of vicinal dicarbonyl compounds with ethanal, hydrogen sulphide and ammonia

Dicarbonyl compounds are often degradation products of sugars. Reaction of vicinal dicarbonyl compounds with ethanal, hydrogen sulphide and ammonia (substance quotients 1:1:2:1) afforded a mixture of thiazolines, thiazoles, oxazolines, oxazoles, imidazoles, 1,3,5-trithiolanes and dithianes.

A survey of these reactions is given in Figure 3.

Alkyl-substituted thiazoles are present in the flavours of tomato (Viani et al., 1969; Wobben et al., 1974), meat (Wilson et al., 1973), peanuts (Walradt et al., 1971), and coffee (Vitzthum & Werkhoff, 1974). 2,4,5-Trimethyl-Δ^3-oxazoline occurs in meat flavour (Chang et al., 1968).

Several alkyl substituted oxazoles have been found recently among flavour components of coffee (Vitzthum & Werkhoff, 1974).

Because of the complexity of the reaction mixtures, we wondered what reactions took place between dicarbonyl compounds and hydrogen sulphide only.

Reaction products of dicarbonyl compounds with hydrogen sulphide

When dicarbonyl compounds were reacted with hydrogen sulphide (substance quotient 1:4) the following reactions took place:

– exchange of oxygen for sulphur with the formation of oxothioketones,

– reduction of the oxothioketones to mercaptoketones, finally then oxidized to the ketoalkyldisulphides,

– formation of 2,5-dialkylthiophenes in the case of 1,4-diketones.

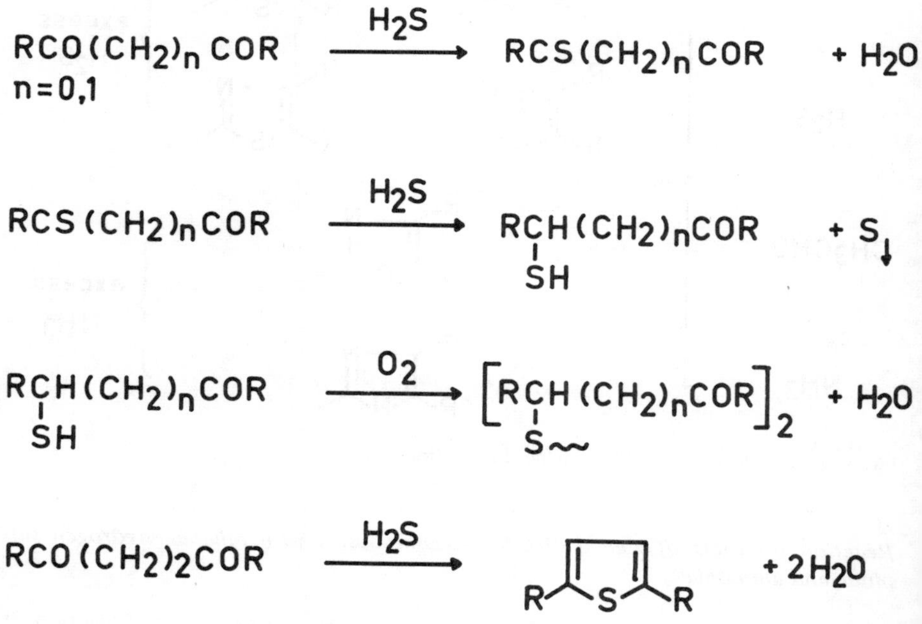

$$RCO(CH_2)_nCOR \xrightarrow{H_2S} RCS(CH_2)_nCOR + H_2O \qquad n = 0,1$$

$$RCS(CH_2)_nCOR \xrightarrow{H_2S} RCH(SH)(CH_2)_nCOR + S\downarrow$$

$$RCH(SH)(CH_2)_nCOR \xrightarrow{O_2} [RCH(S\!\sim)(CH_2)_nCOR]_2 + H_2O$$

$$RCO(CH_2)_2COR \xrightarrow{H_2S} R-C_4H_2S-R + 2H_2O$$

Fig. 4. Reaction products of dicarbonyls with H_2S.

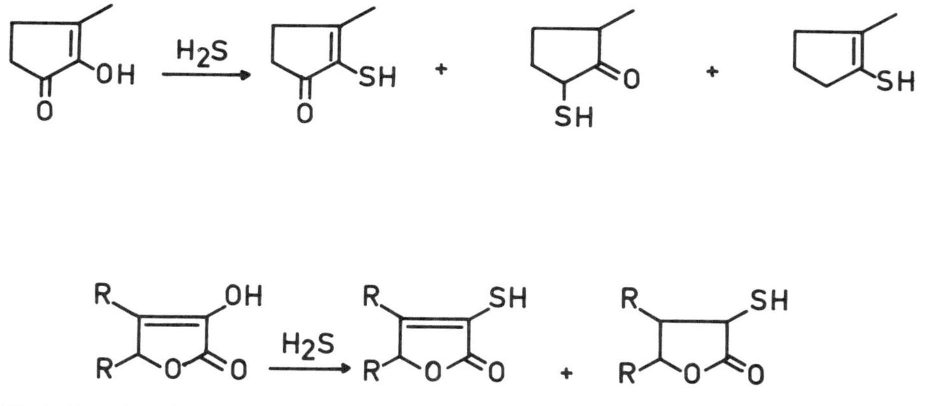

Fig. 5. Reaction of cyclic dicarbonyls with H_2S.

The reaction products of dicarbonyl compounds with hydrogen sulphide are given in Figure 4.

Mercaptoketones can be expected in meat flavour. 2,5-Dialkyl thiophenes occur in the flavours of meat (Wilson et al., 1973), and in onion (Boelens et al., 1971).

As depicted in Figure 5 the same reactions took place when a homocyclic diketone (cyclotene) and an oxygen containing heterocyclic diketone 2,4-dihydroxybut-2-enoic acid lactone (α-tetronic acid) reacted with hydrogen sulphide.

References

Badings, H. T., H. Maarse, R. J. C. Kleipool, A. C. Tas, R. Neeter & M. C. ten Noever de Brauw, 1975. Formation of odorous compounds from hydrogen sulphide and methanethiol, and unsaturated carbonyls. Proc. int. Symp. Aroma Research, Zeist, May 1975. Pudoc, Wageningen, p. 63.

Boelens, M., P. J. de Valois, H. J. Wobben & A. van der Gen, 1971. Volatile flavor compounds from onion. J. agric. Food Chem. 19: 984.

Boelens, M., L. M. van der Linde, P. J. de Valois, J. M. van Dort & H. J. Takken, 1974. Organic sulfur compounds from fatty aldehydes, hydrogen sulfide, thiols and ammonia as flavor constituents. J. agric. Food Chem. 22: 1071.

Brinkman, H. W., H. Copier, J. J. M. de Leuw & S. B. Tjan, 1972. Components contributing to beef flavor. Analysis of headspace volatiles of beef broth. J. agric. Food Chem. 20: 177.

Buttery, R. G., R. M. Seifert & L. C. Ling, 1970. Characterisation of some volatile potato components. J. agric. Food Chem. 18: 538.

Buttery, R. G. & L. C. Ling, 1974. Alkylthiazoles in potatochips. J. agric. Food Chem. 22: 912.

Chang, S. S., C. Hirai, B. R. Reddy, K. O. Herz, A. Kato & G. Sipma, 1968. Isolation and identification of 2,4,5-trimethyl-Δ^3-oxazoline and 3,5-dimethyl-1,2,4-trithiolane in the volatile flavor compounds of boiled beef. Chem. Ind. 1639.

Fujimaki, M., S. Kato & T. Kurata, 1969. Pyrolysis of sulfur-containing amino acids. Agric. biol. Chem. 33: 1144.

Ledl, F., 1975. Analyse eines synthetischen Zwiebelaromas. Z. LebensmUnters. Forsch. 157: 28.

Obata, Y. & H. Tanaka, 1965. Fotolysis of sulfur-containing compounds. Agric. biol. Chem. 29: 191.

Viani, R., J. Bricout, J. P. Marion, F. Müggler-Chavan, D. Reymond & R. H. Egli, 1969. Sur la composition de l'arome de tomate. Helv. chim. Acta 52: 887.
Vitzthum, O. G. & P. Werkhoff, 1974. Neu entdeckte Stickstoffheterocyclen im Kaffee-Aroma. Z. LebensmittelUntersuch. Forsch. 156: 300.
Walradt, J. P., A. O. Pittet, T. E. Kinlin, R. Muraliahara & A. Sanderson, 1971. Volatile components of roasted peanuts. J. agric. Food Chem. 19: 972.
Wilson, R. A., C. J. Mussinan, I. Katz & A. Sanderson, 1973. Isolation and identification of some sulfur chemicals present in pressure-cooked beef. J. agric. Food Chem. 21: 873.
Wobben, H. J., P. J. de Valois, R. ter Heide, H. Boelens & R. Timmer, 1974. Investigation into the composition of a tomato flavour. 4th International Congress of Food Science and Technology, September 1974, Madrid.

27 May – Relation between instrumental and sensorial analysis

Proc. int. Symp. Aroma Research, Zeist, 1975. Pudoc, Wageningen.

The human instrument in sensory analysis

E. P. Köster

Psychological Laboratory, University of Utrecht, Varkenmarkt 2, Utrecht,
the Netherlands

Abstract

In sensory analysis, the human subject is used as an instrument, which registers stimuli from the external world, transforms them into sensations and measures these sensations on a numerical or a descriptive scale.

Since knowledge about the relevant properties of the measuring instrument is prerequisite for good measurement, it seems strange that people working in sensory analysis devote very little attention to the properties of the human instrument. Of all the properties which would be considered extensively in the selection of a physical measuring device, such as sensitivity, reliability, linearity and constant errors, only the first is usually considered in some form or other. Nevertheless, much knowledge is available about the physiological factors that influence the other properties. Thus, good control of what the person expects and of his motivation can improve the reproducibility of his responses considerably.

Other important factors in function are related to memory, selective attention, habituation and sensory adaptation, topics which have been studied extensively by experimental psychologists over the last two decades. Some practical implications of the findings of this research for the optimization of man as an instrument in sensory analysis are discussed.

Introduction

In sensory analysis, man is used as a measuring instrument in an ambiguous way. The ambiguity is mainly because the 'instrument' records its own interactions (or confrontations) with the environment as well as environmental properties. The 'reading' depends on the observer's history and his system ('sense') of values or preferences.

Each observer has his preferences, to some extent determined biologically, but largely acquired by his personal experience. All observers in sensory analysis have a certain capacity and limitation, specific to man and perhaps not present in other species. Thus, we are more sensitive to certain odours than dogs, but dogs are certainly much more sensitive to odours like butyric acid than we are.

In using such a complicated device whose peculiarities obviously play an important role, the investigator should know the properties of his measuring instrument well. However, this is often not so. People with a detailed knowledge of the properties of their physico-chemical instruments may take great care to avoid extraneous variables like temperature shifts during their physico-chemical measurements, but may become careless when using the human instrument in sensory analysis.

Often tests are in rooms whose physical variables, such as lighting, ventilation and humidity, are controlled, but important psychological variables are ignored. Traditionally, the investigator in sensory analysis knows a lot about statistics and little about his instrument. But even the best statistics cannot improve bad measurements and only a thorough knowledge of the instrument can optimize its use.

What do we require of a good instrument and to what extent does a human observer meet these requirements? A good measuring instrument must be sensitive. It must detect small differences in the intensity or the quality of the environment and it must record them in a reliable and reproducible way. If possible, it must be linear or at least it must translate the measured properties of the environment according to a known function. Finally, its systematic errors must be as small as possible and it should be free from unknown systematic errors.

The human instrument meets these requirements only partially. It is certainly sensitive, but its sensitivity varies from person to person. Reliability and reproducibility are rather unsatisfactory. As all workers in sensory analysis know, human observers can easily be influenced. Many external and internal factors influence measurements.

But does this not hold for any measuring instrument? Should one not test these factors and try to control them? This paper will discuss some factors and their interdependence and we will try to indicate ways in which their influence on human measurements may be reduced as much as possible.

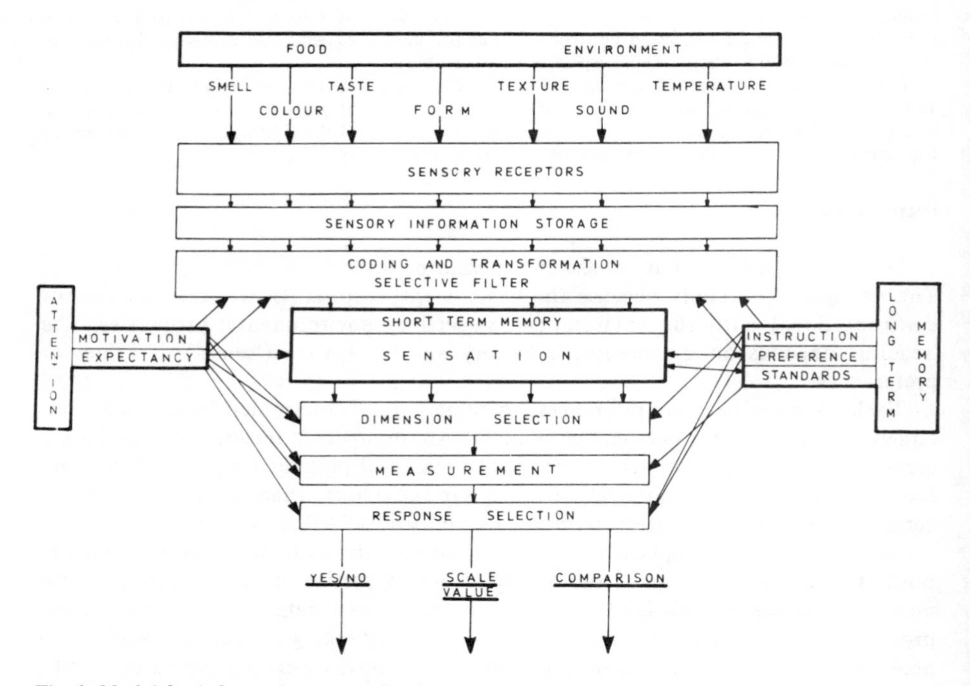

Fig. 1. Model for information processing in sensory analysis.

A schematic model

Over the last few decades experimental psychologists have studied human perception and human information processing. Although knowledge is still inadequate, many findings improve our understanding of human function in sensory analysis. Figure 1 is a scheme of the processes in sensory analysis. The stimuli and the responses by the observer can be linked by intermediate and less easily measurable steps, whose peculiarities and limitations determine the relationship between stimulus and response. The sensory message is recorded, coded, transformed, reduced and measured before response.

Figure 1 shows that the multitude of possible impressions which the stimulus (food and environment) gives is reduced drastically in sensory analysis. Man is a multipurpose instrument and is sensitive to many kinds of stimulus, but he has only a limited processing capacity. Man differs from the most physical instruments, which are only sensitive to a single property. Man's much larger flexibility endangers the validity of measurement. Even if well programmed by verbal or written instruction, the reduction may be incomplete. Everybody knows how difficult it is to judge the smell or the taste of a product independently of its colour or to judge the intensity of an odour without taking into account its nauseating quality. Far too often investigators overestimate the analytic capacities of their panel members. A short instruction like 'Ignore the colour' is considered to be sufficient to exclude it.

But the opposite is found too. Panel members may be asked to complete a long questionnaire after tasting a sample. This seems an economical procedure because repeated testing of the same sample can be avoided. However, it is forgotten that the processing capacity of the human instrument is far too limited. We have difficulty in considering more than one thing at a time. If the panel member is very conscientious he will taste the sample many times and will answer only one or two questions each time he tastes, but more often he will taste only once or twice and cannot then judge each aspect independently. His judgements of certain aspects will depend on his judgements of other aspects.

In general, it is better to let the panel members judge only one aspect of the sample at each presentation.

A second point that prompts itself on inspection of the figure, is man's double task as a measuring instrument. One task is to record and translate stimuli from the outside world into sensations; the other is to measure the sensations and compare them with internal and external standards.

Often it is forgotten that what is measured are not the properties of the stimuli, but rather the sensations the stimuli have produced. These sensations depend also on properties of the recording instrument. Such codeterminant properties include coding and transformation mechanisms, which may be equal in all mankind, and personal history, such as preferences, which vary from person to person. Especially if the investigator tries to relate the human responses to instrumental measurements, it is important not to try to make observers simulate the objective instrument, but to let the objective instrument simulate the human measurement as much as possible. Examples of the opposite reasoning are numerous. In research on meat texture, it is customary to start by developing a gadget that measures a particular physical property of meat and subsequently to look how well the results from human observers correlate with these measurements. It would be better to start by analysing thoroughly the different sensations recorded and measured by man and then to construct measuring devices that imitate measurement of the aspects of the human sensation.

Let us briefly examine two groups of factors which are mentioned in the figure and which influence the stages of the actual process: attention and long-term memory.

Attention is in itself not a part of the sensory analytic activity, but modulates the functioning of some stages of the process. Attention may vary, during the performance of a sensory task and over the longer periods between repetitions of a task (for instance in tests on keepability). Although attention may fluctuate, it can easily be manipulated. The investigator must keep the attention of his panel members as constant (not always as high) as possible. We will discuss the different ways, in which attention, expectancy and motivation influence the process, when we consider the different stages of the process. We will also indicate ways in which fluctuations in attention can be reduced.

Long-term memory also influences many stages. Panel members must bear instructions in mind and they must perform manipulations learned. This memory contains all the standards with which the observer must compare his sensations. Long-term memory plays an important role in the preferences of a panel member. Its influence will be discussed in the same way as attention. But first let us consider the actual stages of sensory perception.

The sensory processes

The senses of smell, taste, touch (hand and tongue), vision, hearing (sounds in the oral cavity) and kinaesthesis (judging hardness, elasticity etc.) are involved in sensory perception. They transduce physical messages from the environment into different neural messages. Often mechanisms in the sense organs themselves (lateral inhibition) accentuate differences in the perceived stimuli so that the difference in the neural message may be larger than in the physical message. On the other hand, very small differences in the physical message may be suppressed in the neural message.

Sensory information storage

All information that is received by the receptors remains available for further processing for a few tenths of a second. What is not processed within that time is lost. If new information is received by the same receptors within that time, the earlier information is usually lost. This brief information storage is the basis for our experience of continuity in perception. Thus, we do not notice when our own eye blinks.

Coding and transformation

Processing the information includes coding and transformation. It is difficult to indicate exactly where in the system the mechanisms for coding and transformation are seated. Some are inclined to place them just before the selective filter. By coding and transformation, neural impulses are translated into 'sensations'. Evidently, this is only a hypothetical step known only by its result. Under certain conditions, the physical intensity (I) of the stimulus is related to sensation evoked (S) according to Stevens' power law: $S = I^n$. Under other conditions, Fechners' law holds: $S = k\log I$. Without going into detail, we can conclude that transformation can be described by general rules. Where the coding of qualitative properties is concerned, the process has to be described by classification.

Selective filter

Since only part of the multitude of impressions can be processed, incoming information must be selected. Even without noticing, we continually select information. When something attracts our attention, other things become less meaningful. The selective filter which decides on accepting or rejecting information, is influenced by a number of factors.

The expectancy of the subject about the likelihood of something happening is an important factor. If it is likely to occur, the information will be perceived more readily. In sensory analysis it is important that the observer knows what he is looking for. Instructions play a role here.

Another important factor is motivation. The subject can reduce or extend the mesh of his filter almost infinitely. In sensory analysis, it is important to keep these influences on the filter as constant as possible within and between sessions.

Short-term memory

Information that has passed the selective filter can be stored for a short period of a few seconds in the 'short-term' memory. If it is not processed within about 6–10 seconds, it is lost unless it is explicitly rehearsed. Thus, a telephone number which we hear, is available for a few seconds, and then we loose it unless we use 'inner speech' to rehearse it. In short-term memory, the messages are stored according to sound or color, but not to meaning. The sound of a word (the phone number) still resounds for a few seconds in our head if we want to write it down after hearing it. In sensory analysis, this short-term memory is important if samples must be compared. A direct comparison of two odour samples is still possible, but comparing three samples successively already leads to difficulties, because people loose impression of the first while smelling the third. Also, one should remember the rather limited duration of the short-term memory in deciding how many questions a panel member can answer after one taste of a sample.

Sensation

In sensory analysis, sensations are measured. These sensations are the end products of the recording of outside stimuli by the senses. However, to a certain extent they also depend upon the personal history of the subject. Preferences acquired in life are important. In some cases, it is the only object of sensory analysis to determine the relationship between the sensation brought about by the sample and the preferences stored in a person's long-term memory. Sensation and memory interact in that the sensations to which the subject is exposed in the experiment strongly influence his pattern of preferences. A foodstuff that was highly appreciated at the beginning of a test may become unbearable after tasting it a few times in succession, or appetite for samples may increase during the test. Internal standards often influence sensation. Sensations may have to be stored as standards in long-term memory. Familiarity with a sensation determines its nature: sensation and long-term memory interact.

Expectation, an aspect of attention, also influences sensation. The intensity of unexpected novel sensations, for instance, is always overestimated. Sensations never appear on an 'uncoloured' backcloth. They are always more or less expected, more or less preferred and familiar.

According to the aim of the measurement in sensory analysis, one must emphasize or minimize the influence of this 'coloured' backcloth. The effect of the sensory test itself on both memory and attention is important. If one wishes to assess the preference for a new product, intended for very frequent use by the consumer, one must familiarize the panel thoroughly with the product before measuring the preference.

In other tests, one must avoid excessive familiarity with the product of the house in order to validate comparison with competing products. This is best achieved by confronting the panel regularly with a variety of products in 'blind' tests. The seeming loss in time resulting from the introduction of samples not to be assessed may be counterbalanced by the loss in novelty when only the familiar product is included in the series of samples.

Sensations are multidimensional, but the dimensions found at this level need not in any way reflect physical properties of the products. Thus the main colours according to which we classify our colour sensations do not reflect the physical continuum of wavelength. Yellow is not more blue than is red although it is nearer to it in wavelength.

Furthermore, dimensions of another sort are involved in the sensations. Impotant psychological dimensions of sensation are: pleasantness – unpleasantness, perceived intensity and the extremeness on the pleasant – unpleasant dimension.

Property selection

Since sensation is multidimensional, whereas only one dimension or aspect of sensation can be measured at one time, the aspect to be measured has to be selected.

Instruction to the subjects is important in this selection. Aspects of attention such as motivation and expectation are also of influence. The properties to be selected have to be related as closely as possible to those of the sensations themselves. In measurement of intensity, this does not provide any problems, but if one just wishes to compare the taste of samples, difficulties may arise. For instance, if the panel has to indicate off-flavours of one component of an end product that is considered unpleasant itself.

Measurement

After selection of the property to be assessed, the actual test takes place. Usually, this is a comparison with internal or external reference. In fact, even if 'absolute' judgments are made, comparison with an internal reference is involved.

The precision of measurement of course largely depends on the subject's motivation.

If detection or discrimination is difficult, the subject has to decide whether or not a signal (or difference in discrimination) is present. As a result of the occurrence of 'neural noise' (spontaneous activity of the neurons) and the overlap in sensation of noise alone and signal plus noise, results may be indecisive.

Response selection

After measurement, the response has to be chosen. This has to be done when the measurement does not lead to an unambiguous conclusion. If so, the subject must gamble on the response and in doing so he will be influenced strongly by his expectations about the probability of occurrence of the signal and by his motivation to give a positive or negative answer. For the type of response that the subject can give, his instruction is of extreme importance. It must not confront him with insoluble problems. Perhaps the greatest difficulty in sensory analysis is to put the right questions. This is all the more difficult, because experimental subjects always answer the questions put to them, no matter how absurd. The fact that the experimenter asks the question gives the panel the impression that it must be answerable. If not, they just guess.

Another important problem in response selection are code and place references. If one codes the samples in a preference test with the letters A, B, C and D, the subjects, whenever they are in doubt will prefer to use Response A. Systematic variation of the codes over the samples presented may provide a solution. This solution can be improved by choosing codes for which the subjects have a less definite preference. For Dutch subjects, we found the letters B, M, T, V and X were of about equal preference.

Place preferences also may be annoying. The middle of three samples is chosen more frequently by a number of subjects. Here also systematic variation is needed.

The most important properties of human assessments

Finally let us survey the most important requirements for a good measuring instrument and see how far human sensory tests meet them.

Sensitivity

In general, the senses are highly sensitive. Olfactory and tactile sense are often more sensitive than physical instruments.

Reliability

The senses are not reliable. Sensitivity may vary with the following:

a. Adaptation Sensitivity varies during stimulation. The intensity and duration of the stimulation are important. The longer and the stronger the stimulation, the more sensitivity is reduced. After stimulation stops, sensitivity recovers, rapidly at first. Nevertheless it may take a minute, at least, before the olfaction, for instance, fully recovers.

b. Physiological condition Sensitivity in women fluctuates with their menstrual cycle. Other physiological variables like diurnal rhythm and body temperature are also of influence.

c. Weather conditions Day-to-day fluctuations in sensitivity may result from the weather. Thus, people are more sensitive to odours the day after a sudden fall in atmospheric pressure. The atmosphere in the test room is of little consequence.

d. Illnesses, colds, etc.

e. Motivation and expectancy This point has been discussed.

Reliability can be enhanced by controlling factors such as time between stimuli (adaptation), signal probability (expectation) and motivation.

Reliability can also be enhanced by using large panels to exclude physiological variables and illnesses and by repeating the measurements to exclude weather conditions.

Linearity

Although the relationship between physical properties of the stimulus and human response to intensity is not linear, it can be predicted (Fechner's Law, Steven's Law).

Systematic error

Subjects show systematic error, such as preferences for codes and places. The effects of systematic error can be eliminated by correction after the test or by randomizing codes and places.

Proc. int. Symp. Aroma Research, Zeist, 1975. Pudoc, Wageningen.

Aroma values – a useful concept?

M. Rothe

Zentralinstitut für Ernährung Potsdam-Rehbrücke der Akademie der Wissenschaften der DDR, 1505 Bergholz – Rehbrücke, Arthur-Scheunert-Allee 114–116, DDR

Abstract

Increasing numbers of aroma compounds identified in foods require ranking for sensory importance. Although in foods character-impact compounds can be expected only in a few cases, there are differences in the contribution of single aroma constituents to flavour. Conclusions from quantitative data of aroma compounds with respect to the aroma quality of the product may be drawn only, if they are assessed in combination with the aroma effectiveness of the substances.

Aroma effectiveness can be measured by means of threshold concentrations. A selection procedure for important aroma compounds represents the determination of 'aroma values', that is the quotient of content and threshold value of a substance. Thus aroma indices simply to be determined can be found as they are needed for solving practical problems in food production.

For reconstruction or imitation of food flavour, however, such data are not sufficient. Threshold values determined in simple model systems cannot represent exactly the medium present in foods. Synergistic effects of various aroma compounds may alter the result under practical conditions. Additionally, aroma values are influenced by lipid content, distribution of aroma compounds between aqueous and lipid phase, and by sorption of aroma molecules to the polymer nutrients in foods.

Introduction

The numbers of aroma compounds being identified in particular foodstuffs characterize the rapid development of flavour research within the last two decades. A comparison of the first list Dr Weurman elaborated in 1963 with those published 10 years later by his colleagues shows that in most foods the number of known aroma components has increased enormously. Figure 1 demonstrates this raising complexity of food aroma with three different products (van Straten & de Vrijer, 1973; Rothe, 1974).

Among these numerous substances, however, key components have been found only occasionally. According to present knowledge, the aroma of most foods depends on simultaneous perception of many components. Consequently, there are many problems in the measurement of aroma with instrumental instead of sensory analysis, in the evaluation of analytical data, and in the development of flavourings.

There are, however, differences in the importance of the single aroma components for total flavour. Hitherto, our knowledge about the most important consti-

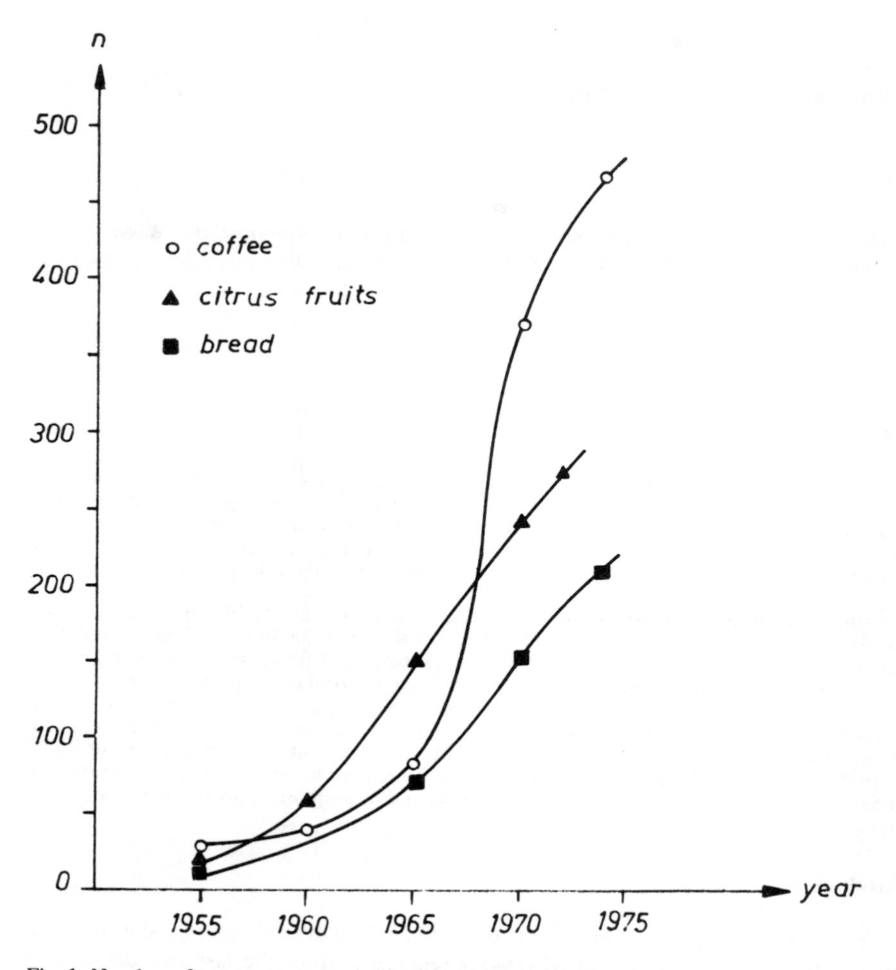

Fig. 1. Number of aroma compounds identified in three foods.

tuents of particular aromas has been limited. Their recognition seems to be one of the key problems in flavour research. Further knowledge would, for instance, facilitate the search for yardsticks of aroma quality or aroma intensity. Better knowledge about predominant substances within the aroma would enable us to reduce the quality difference between natural flavour extracts and synthetic mixtures and would give synthetic mixtures more chance of acceptance by the consumer.

But looking for such components of central importance for total flavour is a difficult problem. Not only the amount of a volatile substance decides its place in the complex of aroma components. Aroma effectiveness seems more important. After failing to find 'character-impact' components among the main volatile substances, we must suppose that minor components of high effectiveness govern a flavour.

The concept of aroma value

To overcome these difficulties, Patton (1957) proposed estimation of sensory threshold values. Our own concept, in studies on bread (Rothe, 1963), we called 'aroma value'. Other authors used the terms 'odour value' (Mulders, 1973) and 'odour unit' (Guadagni et al., 1966). By this concept, two characteristics of an aroma compound may be combined: actual concentration and aroma effectiveness. A measure of 'aroma effectiveness' is threshold value. Since the concept must be as simple as possible for routine use some simplifications are introduced. 'Aroma value' is the quotient of the concentration of an aroma component to its threshold concentration. The value shows how much actual concentration exceeds its threshold. Supposing that substances with equal aroma values have equal odour intensities, we can rank those substances after analysis in a system.

Relation of aroma effectiveness and threshold value

Let me enlarge on aroma effectiveness and threshold concentration. The properties are negatively correlated. The more effective an aroma component the more we can dilute it, with substances like water or air, before the threshold be reached. There is no doubt that thresholds are a measure of aroma effectiveness. In recent years, knowledge has accumulated on thresholds, not only from physiologists but also from flavour scientists. A few years ago nobody would have predicted a difference in magnitude of the threshold, for instance, between acetone and methanethiol of 10^7 (Rothe et al., 1972) or between the unsubstituted pyrazine and its methoxy-isopropyl-derivative of 10^8 (Seifert et al., 1970).

Table 1. Threshold value and 'odour value' of some components present in the head space over wheaten bread (water as medium). Mulders, 1973.

Component	Odour threshold value (mg/kg)	Odour value[1]
Acetaldehyde	0.12	57
Dimethyldisulfide	0.00016	16
Ethanol	900	11
2-Methylpropanal	0.01	10
3-Methylbutanol	0.77	10
Dimethylsulfide	0.001	5
1,1-Diethoxyethane	0.042	4
3-Methylbutanal	0.007	3
2-Methylpropanol	3.2	2

1. Odour values < 1 were found for 2,3-butanedione, propanol, 2-propanone, ethyl formate, ethyl acetate, furan, 2-methylfuran.

Some aroma values

Before assessing the concept of aroma value, at first the following example for such a calculation. The data in Table 1 on flavour of wheaten bread were published by Mulders (1973). The table includes threshold values and 'odour values', the latter calculated as quotient of concentration to threshold concentration.

Mulders (1973) elegantly excluded one source of error in such experiments. A main disadvantage of our work in this direction and of other authors, too, was using threshold values estimated in water, whereas the actual concentration of the aroma components was in the food itself or in an extract. But values from a different system may cause erroneous conclusions. To overcome this problem, Mulders (1973) estimated thresholds in a synthetic aqueous solution with the same head-space chromatogram as the wheaten bread isolates. By this trick, he related the two concentrations to the same system.

Mulders (1973) showed that of the 16 substances tested in the head space over wheaten bread, acetaldehyde was the compound with the highest contribution to the aroma.

Table 1 once again demonstrates that major components must not be the important ones for aroma. Acetaldehyde plays a leading role in flavour of wheaten bread. Some simple alkanols and sulphur compounds seem to contribute to the flavour, too.

Some years ago we found, indeed, that ethanol and the higher alcohols with up to 5 carbon atoms govern the sensoric impression 'fermentation flavour' of wheaten bread (Wölm et al., 1974). This result shows that with the concept of aroma values compounds suitable as 'aroma index' will show up. However, Mulders did not succeed in producing bread-like aroma by combining the substances in the same amounts he found in the head space over bread, so that the concept has limitations.

Aroma value criticized

Mathematical calculations of this type are very speculative. With the concept discussed, simplifications are made which may not be permitted in such a complex sensation like flavour. Two lines of criticism are possible:

- physiological facts and relations are ignored
- neglect of physical factors that can influence aroma perception.

Physiologically the following facts may be criticized:

Aroma values assume that the sensation produced by components is additive. Synergistic or antagonistic effects are ignored because mathematical relations are poorly known.

When calculating and evaluating aroma values, we expect an analogous increase in aroma intensity with concentration above the threshold for different aroma components. In other words, a concentration 50-times the threshold for one aroma component ought to represent the same intensity as an aroma value of 50 for another component. However, various publications demonstrate differences in the increase in intensity with concentration (Köster & Wouters, 1970).

Because such physiological problems have been discussed in detail by Dr Köster (Köster & Wouters, 1973), let us examine some additional aspects of the influence of some physical factors.

Factors influencing threshold value

Many specialists have shown that threshold values, though constant in principle, depend upon several conditions. The main factor is the medium. Using air as the most natural diluting agent is possible only if an olfactometer is available to measure the concentrations being tested. Because only a few laboratories possess such a complicated and expensive equipment, most dilute with water. Water is the main constituent of most foods, so threshold studies in an aqueous medium have some justification.

Guadagni et al. (1963) showed that threshold depends upon the procedure used. The data in Table 2 elucidate this fact. They demonstrate three ways of estimating the threshold of some aroma components dissolved in water. Guadagni used plastic flasks that contained the solutions under test, and which the tester had to squeeze while sniffing. In the sensory test of Mulders the solutions were sniffed in open glasses. In our own procedure, we test the aqueous solution by tasting in the mouth; here the volatile aroma components reach the odour perception area indirectly via the fauces.

Sometimes there are great differences in results between these three methods. Not only different panels but probably also different contitions of evaporation of the compound may be reasons.

We get, however, different thresholds also between media. Differences of up to three magnitudes were found for particular aroma components in a particular test procedure between liquid foods like water, milk, oil, beer or water ethanol mix-

Table 2. Threshold values (mg/kg) of aroma compounds in dependence of the testing procedure (water as medium).

	Guadagni (plastic-flask method)	Mulders (smelling on beakers)	Rothe (tasting in the mouth)
Ethanol	100	900	10
2-Methylpropan-1-ol		3	8
3-Methylbutan-1-ol		0.8	0.3
Butyric acid	0.2		3
Ethyl acetate		0.6	0.1
Acetaldehyde		0.1	0.9
2-Methylpropanal	0.001	0.01	0.001
3-Methylbutanal	0.0002	0.007	0.008
Hexanal	0.005		0.02
Decanal	0.0001		0.0009
2-Phenylacetaldehyde	0.004		0.009
Diacetyl		0.007	0.004
Methane thiol	0.00002		0.00002
Dimethyl sulfide		0.001	0.005
Methional	0.0002		0.00004
Pyrazine	175		300
2-Methoxy-3-isobutylpyrazine	0.000002		0.00001

tures. Let us discuss possible reasons for this phenomenon.

Perception of an aroma component in a liquid or solid system naturally depends on its partial pressure above the system. The extent of volatilization, however, depends not only upon the properties of the volatile substance, but also on the properties of the system. Foods are composed of nutrients in either solid or liquid state. Solids like carbohydrates and proteins can bind low-molecular substances like aroma components in different ways and to different extent. This effect touching our problem of aroma values, too, is discussed in detail under ,'Factors governing the emanation of volatiles from an odorous substrate', p. 141. Liquid components too, like water, oil or water-ethanol mixtures, influence volatility of aroma substances, and thus threshold value. But they operate differently from solids.

Guadagni et al. (1972) estimated thresholds for different media. Table 3 shows thresholds by the plastic flask method of some aroma components in water and odourless vegetable oil. The great differences between the media depended also on the aroma component tested. There are extreme differences in the ratio between threshold values in water and oil. They must be caused by intermolecular relations between the diluting agent and the particular component.

Table 3. Different ratios for odour thresholds in water and vegetable oil. Threshold values from Guadagni, Buttery & Turnbaugh (1972), in parts per 10^9 parts.

Compound	Threshold concentration		Ratio oil to water
	in water	in oil	
Methional	0.2	0.2	1 : 1
2,5-Dimethylpyrazine	1800	2600	1 : 1.4
2-Phenylethanal	4.0	22	1 : 5
Hexanal	4.5	120	1 : 27
Hept-2-enal	13.0	1500	1 : 115
Nonanal	1.0	1000	1 : 1000
Non-2-enal	0.08	150	1 : 1875
Dec-2-enal	0.3	2100	1 : 7000

Table 4. Changes in threshold values influenced by the lipid content of the medium. Threshold values determined by tasting (mg/kg).

	Water	Skim-milk	Milk	Cream
	Lipid content (%)			
	0	0.03	2.5	30
Vanillin	0.02	0.04	0.04	0.05
Diacetyl	0.004	0.007	0.04	0.07
Furfural	0.04	–	0.4	0.7
Maltol	0.6	0.6	0.8	6.0

This observation, however, leads to another conclusion of some practical importance. If there are large differences between thresholds in water and oil, the passing from one phase into the other must alter the perceptibility of the aroma component. There must be consequences not only for aroma intensity of a food, but also for aroma quality.

Such considerations are indeed of practical importance. Table 4 demonstrates with water, milk, and cream for example that increasing lipid content of the system raises the threshold of aroma substances, thus reducing aroma effectiveness. This is the reason why cream ice requires a much higher proportion of flavouring ingredients than simple ices based upon whole milk or non-fat milk.

If the presence of lipids in food systems thus alters aroma properties, the size of the effect must depend also on the distribution coefficients between the aqueous and lipid phases. Because of the heterogeneous structures of aroma compounds, there are large differences in the coefficients. Table 5 includes only a few data on this property. However, it clearly shows that in a combined water/lipid system the main part of some aroma components dissolves in the aqueous phase, thus contributing a strong aroma, whereas others migrate into the lipid phase, losing some of its effective aroma.

For solid foods similar relations must be proposed, but such systems are much more complex. The amount of aroma components volatilized and perceptible will

Table 5. Distribution coefficients of some aroma compounds between water and vegetable oil (equal parts water and oil; 20 °C).

Compound	Ratio water : oil
Ethanol	1 : 0.11
Maltol	1 : 0.13
Diacetyl	1 : 0.45
Butanol	1 : 0.78
Furfural	1 : 1.04
Vanillin	1 : 2.79

Table 6. Perceptibility of aroma compounds in water and wheaten bread. Threshold values determined by tasting (mg/kg).

Compound	Threshold value		Ratio wheaten bread: water
	in water	in wheaten bread	
Diacetyl	0.004	10	2500
Maltol	0.6	800	1330
Ethanol	10	12300	1230
Ethylmaltol	0.08	60	750
Vanillin	0.02	7.5	375
3-Methylbutan-1-ol	0.7	18	26

be influenced by type and strength of sorption to the solids. I will give a few examples by comparison of threshold values in water and wheaten bread. Bread is an appropriate system for this purpose if made without fat and sugar, being then composed mainly of starch, protein, and water.

Table 6 shows the great reduction in aroma perceptibility with bread instead of water. We cannot therefore extrapolate data on threshold concentrations in water to such a solid system without a large error. This criticism applies also to our earlier work in which aroma values were calculated from the thresholds of components in aqueous solutions.

Conclusion

With all the restrictions to appraisal of such aroma values, the concept seems only of limited application. Direct consequences may be drawn only for simple systems. A ranking for the aroma components seems as critical as using only the components of high aroma value for flavouring.

But it is our limited knowledge in this field which restricts the application of aroma value. If we understood more about sensory perception, about synergism and antagonism, and about the influences of the system's properties, we could correct the results of such calculations and broaden the utility of the concept. Without such detailed knowledge, aroma values should be considered with caution.

Nevertheless, large differences in aroma values may be taken as an index of

Table 7. Relative percentage and odour units of aroma compounds in potato-crisp oil. Guadagni, Buttery & Turnbaugh, 1972.

Compound	Relative % in potato-crisp oil	Threshold value (parts/ 10^9 parts of oil)	10^{-3}x Number of odour units
Methional	2.0	0.2	100 000
2-Phenylacetaldehyde	18.0	22	8 180
3-Methylbutanal	5.0	13	3 850
2-Ethyl-3,6-dimethylpyrazine	6.5	24	2 720
Deca-2,4-dienal	7.5	135	560
2-Methylbutanal	7.4	140	530
2-Ethyl-5-methylpyrazine	6.0	320	190
Pent-1-en-3-one	0.1	5.5	180
Hexanal	2.1	120	175
2-Methylpropanal	0.5	43	120
Non-2-enal	1.5	150	100
2,5-Diethylpyrazine	1.0	270	37
2,5-Dimethylpyrazine	6.5	2 600	25
Heptanal	0.6	250	24
Oct-2-enal	0.7	500	14
Hept-2-enal	1.8	1 500	12
Dec-2-enal	1.2	2 100	6
Nonanal	0.1	1 000	1

aroma effectiveness. Before finishing, an example for this expectation. Table 7 shows another result from the interesting work of Guadagni and coworkers on potato-crisps.

The main solvent for aroma compounds in potato-crisps is fat. So Guadagni calculated his 'odour unit' from thresholds in vegetable oil. Methional was ranked first among the aroma compounds he tested. Because of the large difference from other components, we may expect that this substance is dominant in the aroma of potato-crisps. This conclusion is right. In comparisons between whole potato-crisps aroma and 14 components a trained sensoric panel found methional the most similar component.

The use of threshold concentrations for evaluating quantitative aroma analysis today may still raise too many problems of detail. But we must accept that there are only a few other ways of solving the mysterious complexity of flood flavours. We have to learn more about the physiology of aroma perception, about the relations between flavour components, about physical properties that influence aroma effectiveness. If we ignore these problems, the concept of aroma values will fail its purpose. But if we consider it as a tool for increasing our knowledge, there is no doubt that this may be one of several ways of unravelling the puzzle of food aroma.

References

Guadagni, D. G., R. G. Buttery & S. Okano, 1963. J. Sci. Food Agric. 14: 761–765.
Guadagni, D. G., R. G. Buttery & J. G. Turnbaugh, 1972. J. Sci. Food Agric. 23: 1435–1444.
Guadagni, D. G., S. Okano, R. G. Buttery & H. K. Burr, 1966. Food Technol. 20: 166–169.
Köster, E. P. & O. Wouters, 1970. Ernährungs-Umschau 17: 349–354.
Mulders, E. J., 1973. Z. LebensmUnters. Forsch. 151: 310–317.
Patton, S. & D. V. Josephson, 1957. Food Res. 22: 316–318.
Rothe, M., 1974. Handbuch der Aromaforschung, Aroma von Brot, Akademie-Verlag Berlin, p. 11–14.
Rothe, M. & B. Thomas, 1963. Z. LebensmUnters. Forsch. 119: 302–310.
Rothe, M., G. Wölm, L. Tunger & H.-J. Siebert, 1972. Nahrung 16: 483–495.
Seifert, R. M., R. G. Buttery, D. G. Guadagni, D. R. Black & J. G. Harris, 1970. J. agric. Food Chem. 18: 246–249.
Straten, S. van & F. de Vrijer, 1973. TNO Report R 4030, Central Institute for Nutrition and Food Research TNO, Zeist, the Netherlands.
Wölm, G., L. Tunger & M. Rothe, 1974. Nahrung 18: 157–164.

Proc. int. Symp. Aroma Research, Zeist, 1975. Pudoc, Wageningen.

Use of odour thresholds in sensorial testing and comparisons with instrumental analysis

Paula Salo

Research Laboratories of the State Alcohol Monopoly (Alko), Box 350, SF-00101 Helsinki 10, Finland

Abstract

By gas chromatography aroma components of distilled beverages, especially whisky, were identified as mainly volatile carbonyl compounds, fusel alcohols, esters and fatty acids. The odour threshold of an identified aroma component is a yardstick of its contribution to the strength of the aroma perceived. A method was developed for estimating statistically checked odour thresholds. A whisky imitation with 67 ingredient substances and rectified grain spirit was used as a model of the aroma complex. The odour thresholds were used with quantitative gas chromatography to determine the relative odour intensity of individual aroma components, and their mixtures. These relative intensities revealed several possible kinds of interaction in aroma components and their mixtures. By statistics an attempt was made to verify observed synergistic additive or suppressive effects.

Aroma is one of the most important characteristics of alcoholic beverages, but also the most difficult to assess. Gas chromatography shows that the aroma of whisky is composed of several carbonyls, alcohols, fatty acids and esters. The aroma of whisky is a mixed flavour complex, the characteristics being determined by many volatile components and their interactions. The components may not contribute equally to the total aroma. Furthermore, the importance of an individual component in the sensory impression may depend upon interactions with other aroma components.

Aroma composition of whisky

The aroma of whisky consists of more than 200 aroma components (Table 1). Carbonyl compounds with low boiling temperatures and sharp smells are important in aroma despite their low concentration. The unpleasantly pungent portion of the aldehydes appearing in fermentation solutions is removed during distillation. As the beverage matures, remaining aldehydes react with alcohol to form the more pleasantly smelling acetals. The predominant aldehydes in blended whiskies are acetaldehyde, *iso*-butyraldehyde and *iso*-valeraldehyde. Furfural, with its grain-like aroma, has also been found in whisky. Diacetyl and 2,3-pentanedione are the most important vicinal diketones (Ronkainen & Suomalainen, 1969).

Fusel alcohols form the predominant group of components in whisky aroma. These alcohols are formed from sugars and amino acids during fermentation. Their

Table 1. Number of aroma components in whisky (Suomalainen & Nykänen, 1972).

	Number of components	Total
Aliphatic carbonyl compounds	17	
Cyclic and aromatic carbonyl compounds	6	34
Acetals	11	
Alcohols	25	25
Monocarboxylic acids	25	
Aliphatic dicarboxylic acids	3	32
Aromatic carboxylic acids	3	
Hydroxy acids	1	
Lactones	3	3
Monocarboxylic esters	59	
Dicarboxylic esters	4	69
Aromatic carboxylic esters	6	
Phenolic compounds	22	22
Hydrocarbons	9	9
Nitrogen compounds	12	12
Sulphur compounds	7	7
Sugars	4	4
Unclassified compounds	9	9
Total		226

composition is also influenced by the distillation method. 3-Methylbutan-1-ol, *iso*-butyl alcohol, 2-methylbutan-1-ol and *n*-propyl alcohol are predominant. Not only the composition of the fusel fraction but also its concentration varies from one type of whisky to another (Suomalainen, 1971).

The free fatty acids in fermented solutions distil easily with steam and alcohol and thus appear in appreciable concentrations also in other distilled beverages than whisky. Besides the short-chain fatty acids from acetic acid to *iso*-valeric acid, long-chain fatty acids with even numbers of carbon atoms are major components. Palmitoleic acid is an important unsaturated fatty acid in Scotch whisky. The major acid component is acetic acid (Nykänen et al., 1968).

The esters and other neutral compounds are numerically the largest group in distilled beverages. Because esters generally have a pleasant and, some of them, a very intense odour, it may be assumed that they appear as important aroma components. 3-Methylbutyl, *iso*-butyl and ethyl esters corresponding to the acids in the whisky, some alcohols of high boiling temperature and at least one lactone were found in the aroma fraction of a Scotch malt whisky (Fig. 1). The typical strong stearin-like smell of Scotch malt whisky is thought to be due to long-chain alcohols and esters of long-chain acids, especially ethyl palmitate and palmitoleate (Suomalainen & Nykänen, 1970a).

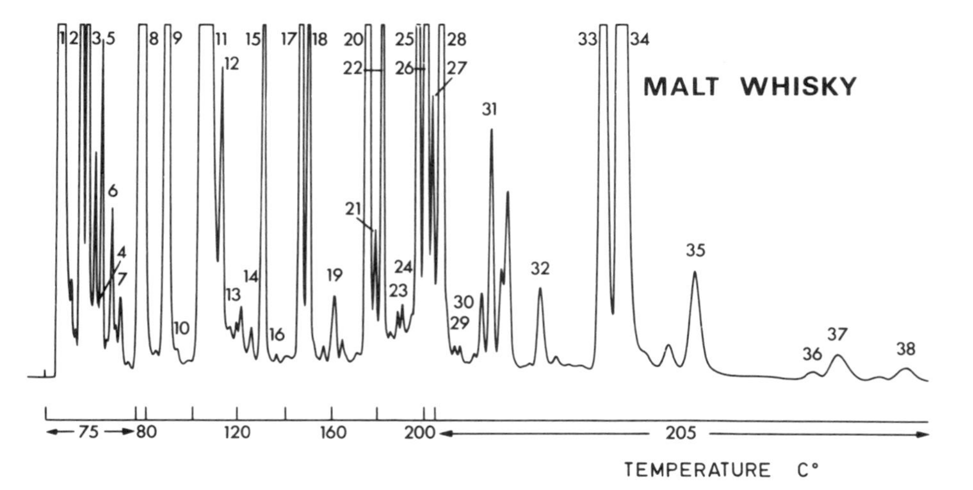

Fig. 1. Gas chromatogram of the aroma compounds of Scotch malt whisky. 1 solvent, 2 ethyl acetate, 3 ethanol, 4 ethyl *iso*-butyrate, 5 unknown, 6 *iso*-butyl acetate, 7 ethyl butyrate, 8 *iso*-butyl alcohol, 9 3-methylbutyl acetate, 10 1-butanol, 11 3-methylbutan-1-ol, 12 ethyl hexanoate, 13 3-methylbutyl *iso*-butyrate, 14 3-methylbutyl *iso*-valerate, 15 1-hexanol and ethyl lactate, 16 3-methylbutyl valerate, 17 ethyl octanoate, 18 3-methylbutyl hexanoate and furfural, 19 1-octanol and ethyl nonanoate, 20 ethyl decanoate, 21 3-methylbutyl octanoate, 22 ethyl 9-decenoate, 23 1-decanol and ethyl undecanoate, 24 β-phenethyl formate, 25 β-phenethyl acetate, 26 ethyl dodecanoate, 27 3-methylbutyl decanoate, 28 β-phenethyl alcohol, 29 1-dodecanol, 30 2-methyl-4-hydroxy-octanoic acid lactone, 31 ethyl tetradecanoate, 32 1-tetradecanol, 33 ethyl hexadecanoate, 34 ethyl palmitoleate, 35 1-hexadecanol, 36 ethyl octadecanoate, 37 ethyl oleate, 38 ethyl linoleate (Suomalainen & Nykänen, 1970b).

Although the aroma of whisky is composed of hundreds of aroma components, the same main volatiles appear not only in the aroma fraction of different types of whisky but also in the aroma of beer, wine and other distilled beverages (Suomalainen & Nykänen, 1972; Suomalainen et al., 1974). Thus the aroma composition seems to depend only slightly on the nature of the raw materials. The quantitatively most important aroma components are produced by the yeast during fermentation. In distilled beverages, the aroma composition is further affected by the technique of distillation and by maturing.

In spite of the similarity of the aromas by chemical analysis, it is easy to differentiate between one beverage and another by flavour. As the number of identified aroma components increases, it becomes more difficult to determine which contribute most to the characteristic aroma of an alcoholic beverage. As yet, aroma composition cannot be expressed by analytical data alone; in other words, such data do not characterize the quality of the beverage in the same way as does an organoleptic test. This is probably because the aroma is sensed as a whole of all the components whereas the chemical analysis only reveals individual components and their concentrations.

Thresholds

Aroma is the sensory response evoked by volatiles. The question arises whether each component exerts an influence in the aroma equivalent to its concentration in a beverage. One means of estimating the contribution of an individual compound to the total aroma is by determining its threshold. Thresholds are commonly used as a measure of the flavour significance of aroma substances (Harrison, 1967; Rothe et al., 1972; Meilgaard et al., 1970). Before a threshold can be considered characteristic of a sensory system, it has to be specified in statistical terms.

A method was developed for determining the perception thresholds of some volatiles detected in whisky by gas chromatography (Salo, 1970a). The panel was selected from the laboratory staff. A triangular test was applied, a solution of pure grain spirit in water, mass fraction of ethanol 0.094, being used as solvent and for blanks. Because individuals vary in their response to the odour of pure compounds and even more to a mixture, the number of odd samples correctly identified in the triangle test were corrected by using the percentage-above-chance scores. When the concentrations were plotted against these scores on log-probability paper, the graphical model seemed to fit a straight line (Fig. 2). The regression line of the percentage-above-chance scores proved to be a useful model, and the distribution of the scores could be assumed to follow the normal probability function.

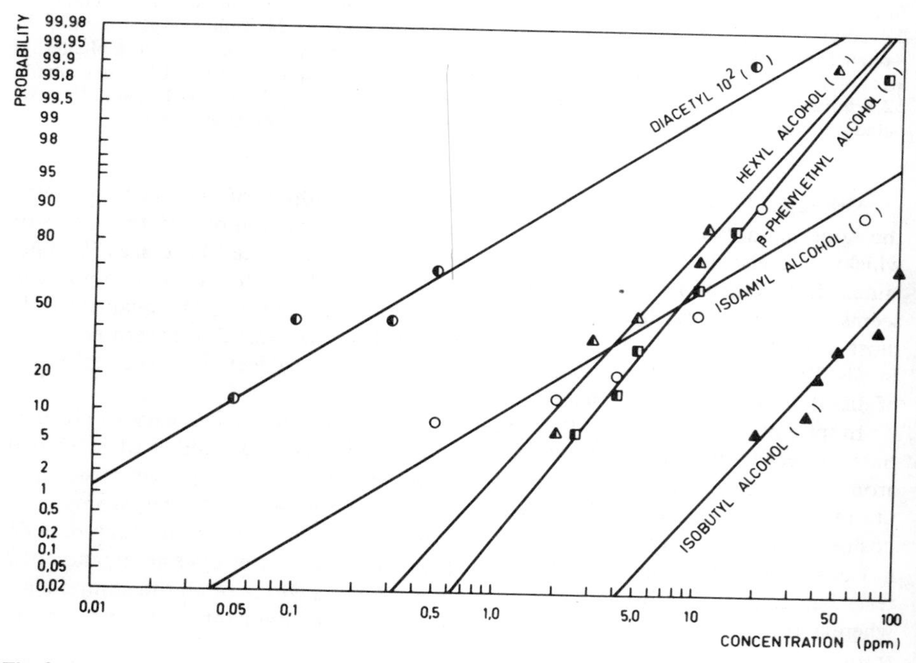

Fig. 2. Percentage-above-chance scores as a normal probability function of logarithmic mass concentration (ppm) of some alcohols and diacetyl (Salo, 1970a).

Table 2. Threshold dilutions of 9 whiskies of various types and of a whisky imitation (Salo, unpublished).

	Threshold dilution x 10^{-4}	One standard deviation range x 10^{-4}
Scotch malt whisky	0.56	0.2 –1.3
Scotch, old blended whisky	0.87	0.5 –1.5
Scotch blended whisky	1.20	0.3 –4.2
Irish whisky	1.30	0.4 –2.0
Scotch whisky, bottled in Finland	1.40	0.6 –3.3
Scotch whisky, blended with Finnish grain spirit	2.00	1.0 –10.0
Bourbon whisky	2.40	0.6 –11.5
Irish whisky	4.50	2.7 –7.5
Artificial whisky imitation	8.80	2.0 –39.0
Canadian whisky	10.40	3.0 –37.0

The concentration that, on average, is detectable in 50% of the cases is commonly accepted as the threshold. At first the threshold values were read from the intersection of the regression line with the 50 percent-above-chance score ordinate. Subsequently, a special computer program was developed, in which the thresholds and their standard deviations could be read directly. With this method, the threshold values were mainly dependent upon the substances themselves (Salo, 1970b).

The detection threshold could be estimated, not only for individual compounds but also for strength of the total aroma of a beverage. The identification threshold was determined by diluting a sample of whisky with pure distilled water until the characteristic aroma of whisky was just recognized (Table 2).

A whisky imitation

The thresholds merely represent the situation when a component is alone in a system, here a solution of rectified grain spirit and water. The aroma of an alcoholic beverage, however, is typically a mixture of many components, and the concentration and proportions of these components can vary within limits without any considerable change in perceived aroma (Suomalainen & Nykänen, 1970b). Clearly, more attention is needed to the proportions of these aroma components and their sensory effects.

An artificial beverage, an imitation of blended Scotch whisky, was prepared in accordance with data from gas chromatography (Salo et al., 1972). The whisky imitation consisted of 9 carbonyls, 13 alcohols, 21 acids, 24 esters and highly rectified grain spirit. When the imitation was mixed with a Scotch whisky, the panel could detect no significant change in whisky aroma until the mixture contained at least 60% of the imitation beverage. It was assumed that this synthetic whisky model reflected the composition of the aroma in genuine whiskies and could be used as a model of the aroma complex that actually occurs in whiskies.

In addition to thresholds for individual compounds, thresholds were also determined for some mixtures of compounds, for aroma fractions of alcohols, acids,

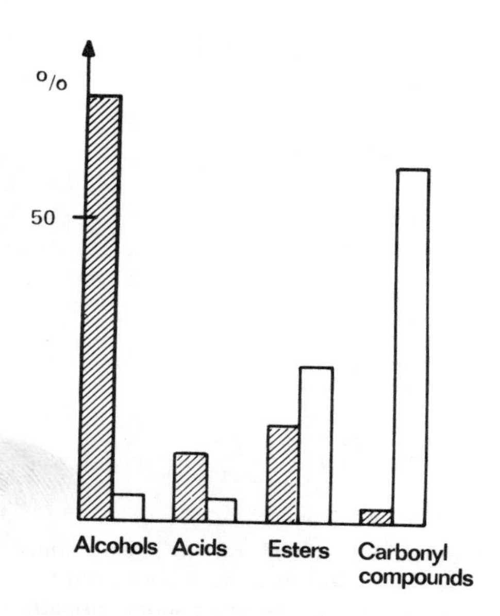

Fig. 3. Mass fraction (%) (hatched) of alcohols, acids, esters and carbonyl compounds in total system and their contributions (%) to 'odour units' (white) of a whisky imitation (Salo, unpublished).

esters and carbonyls, and even for the whole beverage. The relative contribution of each component, mixture or fraction to the odour was estimated by dividing the concentration by the respective threshold. This quotient was called the *'odour unit'*. The contribution of fractions of alcohols, esters, fatty acids and carbonyls differs markedly, according to whether they are measured in odour units or relative mass (Fig. 3).

Odour interactions

It proved difficult to express the relative intensity of aroma fractions and mixtures containing several components as a sum of the odour units of the separate components. Furthermore, although the tests were under controlled conditions, the estimates of threshold were variable. Consequently a standard deviation for the threshold values was estimated. At the threshold of a mixture, its odour was just perceptible, and the sum of the odour unit values of the components was equal to the odour units of the mixture, assuming components did not interact (Salo, 1973).

$$O_B + O_C + O_D \ldots = O_1$$

where

$O_B, O_C, O_D \ldots$ = odour unit values of the components B, C, D ...
O_1 = odour unit value of a Mixture 1.

Division of both sides of the equation by the odour unit value of the mixture (O_1) gives the predicted contribution of each component to the odour:

$$100 \times \left[\frac{O_B}{O_1} + \frac{O_C}{O_1} + \frac{O_D}{O_1} + \ldots \right] = 100$$

where:
B, C, D . . . = components in the mixture
1 = Mixture 1.

If the observed sum of the *proportion values* ($100 \times \frac{O_X}{O_1}$) of the components in a mixture equals 100, the contributions to aroma are additive. If the sum is less than 100, the components are synergistic, and if the sum exceeds 100, the components interact suppressively. The standard deviation of the proportion values could be calculated from the threshold values and standard deviations of the individual components and the threshold value and standard deviation of the mixture. Furthermore, we tested statistically whether the calculated total deviated significantly from the expected value of 100.

Some examples of interactions among aroma components in the whisky imitation are shown in Tables 3–5. In the alcohol fraction (Table 3), only three alcohols have an odour unit more than one, and all alcohols seemed additive. Nevertheless, the proportion value of 3-methylbutan-1-ol did not deviate significantly from the proportion value of the whole fraction. Thus, 3-methylbutan-1-ol dominates the aroma intensity of the alcohol fraction. 3-Methylbutan-1-ol has also been observed to be the major higher aliphatic alcohol in beer (Meilgaard et al., 1971). Alcohol Mixture 2 represents a model mixture in which the two components contribute equally to the total aroma.

Table 3. Contribution of alcohols to strength of aroma of the alcohol fraction of the whisky imitation (Salo et al., 1973).

	'Odour unit'		'Proportion value'	
	in Mixture	in alcohol fraction	in Mixture	in alcohol fraction
Alcohol Mixture 1 (9 alcohols with odour units less than 0.5)		2.1		4.7
Alcohol Mixture 2		5.0	103.0^{ns}	11.0
iso-Butyl alcohol	2.3		46.0	
2-Methylbutan-1-ol	2.9		57.0	
3-Methylbutan-1-ol		36.0		78.6
β-Phenethyl alcohol		0.6		1.3
Total alcohols		45.4		95.6^{ns}

ns = Total value does not deviate significantly from the expected value 100 ($P > 0.10$).

Table 4. Contribution of acids to the strength of aroma of the acid fraction of the whisky imitation (Salo et al., 1972; Salo, 1973).

	'Odour unit'		'Proportion value'	
	in Mixture	in acid fraction	in Mixture	in acid fraction
Acid Mixture 1 (13 acids with odour units less than 0.1)		0.8		1.9
Acid Mixture 2		3.3	24.6***	8.4
iso-Butyric acid	0.3		9.6	
Butyric acid	0.2		7.2	
Hexanoic acid	0.3		7.8	
Acid Mixture 3		57.4	16.6***	144.0
Acetic acid	1.4		2.3	
iso-Valeric acid	4.3		7.4	
Octanoic acid	0.9		1.6	
Decanoic acid	1.8		3.2	
Dodecanoic acid	1.2		2.1	
Total acids		39.8		154.0**

** = Total deviates significantly from the expected value of 100 ($P < 0.01$).
*** = Total deviates highly significantly from the expected value of 100 ($P < 0.001$).

In the acid fraction (Table 4), Acid Mixture 2 and Acid Mixture 3 both display synergism, although in Mixture 2 all components were present at subthreshold concentrations. However, suppressive effects were displayed in the total acid fraction. Because of its lower threshold than other acids, *iso*-valeric acid may lower the threshold of Acid Mixture 3, but could not affect the threshold of the whole fraction to the same extent. This may explain the suppressive effects found, and suggests how profoundly the aroma is influenced by even small variations in the amount of *iso*-valeric acid.

Synergism was observed in the total strength of the aroma of the whisky imitation (Table 5). Carbonyl compounds and esters should make a larger contribution to the total aroma than do alcohols and acids. The combined contribution of alcohols and acids amounts to less than a tenth of the total odour, although their mass fraction in total aroma fraction is fully 0.8 (Fig. 3). 3-Methylbutan-1-ol, and the aroma components of Acid Mixture 3, Ester Mixture 3 and Carbonyl Mixture 2 contribute substantially to the aroma intensity of this whisky imitation. Additivity replaced synergism when the interactions within aroma fractions were included in the calculation for the total aroma, so that the interactions that influence the perceived total aroma can perhaps be calculated by the contributions of individual aroma components.

Table 5. The aroma intensity of the whisky imitation (Salo et al., 1972; Salo 1973).

	'Odour unit'	'Proportion value'
Alcohol Mixture 1 (Table 3)	2.1	0.2
Alcohol Mixture 2 (Table 3)	5.0	0.4
3-Methylbutan-1-ol	36.0	3.2
β-Phenethyl alcohol	0.6	0.05
Alcohols	45.4	4.0
Acid Mixture 1 (Table 4)	0.8	0.07
Acid Mixture 2 (Table 4)	3.3	0.3
Acid Mixture 3 (Table 4)	57.4	5.1
Acids	39.8	3.5
Ester Mixture 1	12.1	1.1
Ester Mixture 2	8.9	0.8
Ester Mixture 3	137	12.1
Esters	296	26.2
Carbonyl Mixture 1	18.0	1.6
Carbonyl Mixture 2	542	48.0
Carbonyl compounds	658	58.2
Whisky imitation total	1130	72.9*** from Mixtures 91.9^{ns} from fractions

ns = Total does not deviate significantly from the expected value of 100 ($P > 0.10$).
*** = Total deviates highly significantly from the expected value of 100 ($P < 0.001$).

References

Harrison, G. A. F., 1970. The flavour of beer – A review. J. Inst. Brew., London 76: 486–495.

Meilgaard, M., A. Elizondo & A. McKinney, 1971. Beer flavor characterization. Panel testing and gas chromatography for evaluation of thresholds and interactions of some flavor compounds. Wallerstein Lab. Commun. 34: 95–109.

Meilgaard, M., A. Elizondo & E. Moya, 1970. A study of carbonyl compounds in beer, Part 2. Flavor and flavor threshold of aldehydes and ketones added to beer. Tech. Quart., Master Brew Ass. Amer. 7: 143–149.

Nykänen, L., E. Puputti & H. Suomalainen, 1968. Volatile fatty acids in some brands of whisky, cognac and rum. J. Food Sci. 33: 88–92.

Ronkainen, P. & H. Suomalainen, 1969. Die Bestimmung vicinaler Diketone in Weissweinen unter Anwendung der Dampfraum–Analyse. Mitt. Hoeheren Bundeslehr. Versuchsanst. Wein-Obstgartenbau, Klosterneuburg 19: 102–108.

Rothe, M., G. Wölm, L. Tunger & H. -J. Siebert, 1972. Schwellenkonzentrationen von Aromastoffen und ihre Nutzung zur Auswertung von Aromaanalysen. Nahrung 16: 483–495.

Salo, P., 1970a. Determining the odor thresholds for some compounds in alcoholic beverages. J. Food Sci. 35: 95–99.

Salo, P., 1970b. Variability of odour thresholds for some compounds in alcoholic beverages. J. Sci. Food Agric. 21: 597–600.

Salo, P., L. Nykänen & H. Suomalainen, 1972. Odor thresholds and relative intensities of volatile aroma components in an artificial beverage imitating whisky. J. Food Sci. 37: 394–398.

Salo, P., 1973. Interactions among volatile aroma compounds in a whisky imitation. Lebensm. Wiss. Technol. 6(2): 52–58.
Suomalainen, H., 1971. Yeast and its effect on the flavour of alcoholic beverages. J. Inst. Brew., London 77: 164–177.
Suomalainen, H. & L. Nykänen, 1970a. Investigations on the aroma of alcoholic beverages. Naeringsmiddelindustrien 23: 15–30.
Suomalainen, H. & L. Nykänen, 1970b. Composition of whisky flavor. Process Biochem. 5(7): 13–18.
Suomalainen, H. & L. Nykänen, 1972. Formation of aroma compounds in alcoholic beverages. Wallerstein Lab. Commun. 35: 185–202.
Suomalainen, H., L. Nykänen & K. Eriksson, 1974. Composition and consumption of alcoholic beverages – A review. Amer. J. Enol. Vitic. 25: 179–187.

Proc. int. Symp. Aroma Research, Zeist, 1975. Pudoc, Wageningen.

Techniques for assessing odour: uses and limitations

D. G. Land

Agricultural Research Council, Food Research Institute, Colney Lane, Norwich NR4 7UA, England

Abstract

The main methods which are available for characterization and measurement of odour qualities – free description, flavour profile, directed description with rating and semantic differential – are briefly outlined. Each method is analysed as far as available data allow, in terms of requirements for personnel, including training, sensitivity or power of discrimination, reproducibility and the interpretation and limitations of the data obtained. Some effects of variations between people (both individually and as panels), between stimuli and between different qualities perceived on the results are discussed. Finally, some potential areas of development and improvements of techniques are suggested.

Introduction and review

Characterization or description of odour has been an activity of man from his earliest existence. It arose from his primitive need to communicate information on odour, relating to food, the hunt, or even his own safety. Despite great advances in our understanding of the chemical stimuli (for all odours are the result of a chemical stimulus) and investigation of the qualities required for adequate description, odour is still largely described by object language, i.e. in terms of its source, e.g. mushroom. Such terms themselves allow for much variety of qualification, e.g. which species of mushroom? Raw or cooked? Immature, mature or decomposing? In the search for more precise and discriminating ways of characterizing odours, many attempts have been made to characterize them in terms of multiple-constituent odour qualities or notes. These range from methods based entirely on matching to chemical standards (Crocker & Henderson, 1927; Schutz, 1964; Amoore & Venstrom, 1966) which are not now used in practice to those in which affective (feeling or emotional) terms without standards are used (e.g. Paukner, 1965). All are basically profile methods, in which odours are expressed in terms of relative amounts of, or similarities to, constituent characters.

Methods in use fall into four partially overlapping techniques. These are *semantic differential, free description, 'flavour profile'* and *directed description.*

The method of semantic differential has been applied to odour by Paukner (1965) and Yoshida (1964), and consists of rating the odour in terms of a list of uni- or bipolar descriptions relating to attitudes or feelings. It requires at least

20 people and has been applied more with consumer research in mind than as a laboratory procedure for describing odours.

Free description methods consist of the unstructured assessment of qualities. One asks no question other than 'What are the constituent notes of this odour?', usually in the order they are perceived. The method is used extensively by professional perfumers, and wine and tea tasters. Performance with experts is generally thought to be good but in at least one study (Jones, 1968) some doubt has been cast upon their performance. When inexperienced, but science-orientated subjects were tested in groups of about 50, many terms were used, although with most stimuli there was a clear indication of major qualities (Harper et al., 1968a); however a wide range of other terms were also used by a small proportion of subjects. The method provides useful supplementary information in conjunction with difference tests and also provides a basis for the extensively used flavour-profile technique originated by the Arthur D. Little organization.

The flavour profile method involves 4–5 highly trained assessors, one of whom is the leader. The panel evaluates the constituent notes of an odour, their approximate strength, the order of appearance and the overall impression first independently and then by discussion, and arrives at a consensus evaluation. The method has been extensively used but little data on its performance have been published although it is claimed to reproduce consistently (Caul, 1957). The technique hardly requires further description.

The directed descriptive approach is a development of the flavour profile but differs in that a series of specific questions on how much, if any, of each of an extensive list of odour characters is present. The results, which are obtained by each of about 10–12 assessors independently, can be assessed by statistical methods. It is based largely on the technique of Pilgrim & Schutz (1957) and was extended in the work of Harper et al. (1968a, b), applied to some stimuli proposed as odour standards (Land et al., 1970), and first applied to a practical food problem (bilberry juice) by von Sydow et al. (1970). The method has since been developed further and used for blackcurrants (von Sydow & Karlsson, 1971) and to cooked meats (Persson et al., 1973). The method has also been used by Clapperton (1973, 1974) on beers and a similar method using an interval scale (Stone et al., 1974) has been applied to beer (McCredy et al., 1974).

Directed odour description

The technique involves using a panel of selected assessors to rate for intensity all relevant qualities of an odour as a series of successive questions. Ideally the panel is selected from people of known consistency, discrimination and sensitivity who have a wide background of odorous chemicals and the ability to express the character in words. In practice it is not difficult in a laboratory to obtain enough people who perform adequately with further special training.

The test form is evolved from a generalized form (Harper et al., 1968a) supplemented with any additional terms required. 'Round-table' sessions with specific samples result either in clarification of the use of divergent qualities or in agreement to differ; if there is disagreement the terms are retained.

Training is normally carried out with a range of samples of the material to be tested. Repeated assessment of the same materials provides an index of improvement in consistency. It is difficult to make specific recommendations on training as the replication necessary will be greatly influenced by the previous experience of the panel with the product or similar products, and with the use of other sensory descriptive techniques.

Panel performance

Performance of a panel is a composite of consistency (ability to replicate with little variance), discrimination (ability to distinguish between small differences in qualities) and sensitivity (ability to detect traces). Clearly these factors are interrelated; for example if a panel could not discriminate differences but was consistent, the results would be valueless. A sensitive panel may be less consistent than a less sensitive panel which cannot detect such small differences; however the results from the former may be very useful despite the greater variance. Clearly the ideal panel would be sensitive, discriminating and consistent. The problem of how to measure and assess these is far from simple.

Any panel is composed of individuals, and, as living organisms, will have inherent biological variation. Consequently they will show both within-individual and between-individual variation. The former can occur with some otherwise normally consistent individuals, who may have an 'off-day'. Such variations may only occur occasionally but may reflect physiological changes. Differences between individuals are well-known, particularly in preference. The saying 'One man's meat is another man's poison' occurs in different forms in many cultures. Such differences may be much more fundamental than preference or conditioned response – the data in Table 1 suggest a clear difference in the character perceived. Although at present these variations, which are quite real and consistent, can only be observed and noted, in the future it may be possible to use the information much more fully in predicting minority population responses.

Table 1. Individual variation in odour perception by experienced subjects for phenylacetic acid (1.5% in dinonyl phthalate). Qualities and intensity ratings on 0–5 scale. Hedonic rating on 9-point scale.

Subject 2		Subject 17	
musk like	3	putrid	4
floral	3	animal	4
sweet	2	sweaty	3
fragrant	2	faecal	2
sharp, pungent	2	meaty (cooked)	1
aromatic	1	burnt, smoky	1
heavy	1	ammoniacal	1
almond	1	oily, fatty	1
very pleasant	2	sharp, pungent	1
strength	3	very unpleasant	8
		strength	4

Table 2. Within-subject range in replicate ratings. Individual odour character notes on a 0–5 scale. Cumulative % frequency distribution.

Range between replicates	10 substances (2 replications panel size: 10)	3 substances (3 replications panel size: 8)	5 bilberry juices (4 replications panel size: 10)
00[1]	24	12	19
0+[1]	42	24	34
0–1	74	61	72
0–2	91	90	92
0–3	97	97	98
0–4	100	97	99
0–5	100	100	100

1. 00 = qualities consistently absent
   0+ = qualities consistently present

Table 3. Odour qualities of bilberry juice (raw data). 0–5 rating scale.

Quality	Assessor										Total
	C	G	H	B	D	J	R	T	N	M	
Fruity	0	2	3	2	2	3	3	3	5	5	28
Green, cut grass	2	1	2	4	1	1	0	1	4	1	17
Sweet	0	2	1	1	1	0	3	1	2	4	15
Apple-like, raw	0	0	2	1	1	2	2	2	0	5	15
Blackberry	0	1	1	0	2	0	2	0	4	2	12
Fragrant	0	2	1	0	0	0	3	1	1	2	10
Strawberry, unripe	0	3	1	0	0	0	0	1	0	3	8
Apple-like, cooked	2	0	0	0	0	0	0	0	3	2	7
Sickly	2	1	1	0	0	0	0	1	1	1	7
Sharp, pungent	0	0	2	0	1	1	0	1	2	0	7
Blackcurrant	0	1	1	0	0	2	0	2	0	0	6
Metallic	2	0	0	0	1	0	1	0	1	0	5
Strength	3	3	4	3	2	4	2	3	3	4	
Total quality ratings	8	13	15	8	9	9	14	13	23	25	

The extent to which these differences in response between people are considered in selecting a panel depends upon the purpose of the work. If the purpose is to characterize the sensory properties of a material, a panel is required which is not selected other than for their ability to use the technique consistently, and for general discrimination of odours. On the other hand, if the purpose is to follow changes in a particular product, for example in order to optimize processing conditions for the best sensory quality, then a panel which represents the more sensitive part of the population to variations in that product is required, and selection on discrimination and sensitivity to the attributes of that particular product is required in addition to consistency and general discrimination.

In practice this variation between people seems to account for far more variability than that within people. Consistency of response of individuals is often remarkable when the difficulty of the task is considered. Estimates of odour strength of 10 chemicals in duplicate by 10 people and of three chemicals in triplicate by 8 people showed that 94% and 88%, respectively, of responses agreed within one unit on a 0–5 scale where in both cases the mean rating was 3.1. Similarly with individual qualities (Table 2) the same stimuli were assessed with 90% of replicate values within a range of 2 units (i.e. ± 1); similar results were also obtained with more complex stimuli (bilberry juice).

When each individual assesses an odour consistently, then clearly the panel as a whole will also be consistent. Table 3 illustrates some raw data from one of five assessments of bilberry juice. Major qualities were found by most people but none by all. Some minor qualities were found to be important to a few people. The quality 'apple-like, cooked' was scored by the same three assessors 14 out of 15 times in five replicates. Similarly the four who used 'strawberry-like, unripe' used it in each of the replicates. Similar minority usage of certain terms can also be seen in the data of Persson et al. (1973) and of Clapperton (1973). Table 3 also illustrates the wide range in number habit between people. Some use few qualities and low numbers; others use many qualities and high numbers, but they are consistent. At the Food Research Institute we have never tried to 'train' people into greater uniformity of number use lest the change in established number habits increases variability. These results also reflect variation in sensitivity between individuals.

Aggregate panel consistency is shown in results of five replicate assessments of bilberry juice (Table 4) (von Sydow et al., 1970), where the standard errors of the means are low. A similar order of panel variance was found by Persson et al. (1973) on meats.

A further illustration of panel consistency and stability with time is shown in Table 5. On three occasions over a period of 15 months, a series of chemical stimuli was assessed. Results from the whole panels and also from the eight people com-

Table 4. Major odour qualities of bilberry juices. Mean scores, ranges and standard errors of 5 replications.
Comparison of mean scores of 2 juices with 1 and 3 μg/l ethyl 2-methylbutyrate, respectively.

Odour quality	Bilberry juice (5 replicates)			Stripped bilberry juice + *trans*-2-hexenal (3 μg/l) + ethyl 3-methylbutyrate (0.06 μg/l) + ethyl 2-methylbutyrate (1 or 3 μg/1)	
	mean panel score	range	standard error of the mean	1 μg/l	3 μg/l
Fruity	25.6	± 2.3	± 0.9	24.0	25.3
Green, cut grass	17.4	± 1.9	± 0.7	14.5	14.7
Sweet	12.6	± 2.0	± 1.0	13.5	14.3
Apple-like, raw	12.4	± 2.5	± 1.8	15.5	18.3
Blackberry	12.0	± 1.4	± 0.5	7.5	12.1
Fragrant	8.4	± 1.5	± 0.8	6.5	8.9
Apple-like, cooked	8.2	± 1.9	± 1.1	9.5	8.5
Strength				2.65	2.70

Table 5. Effect of change in panel personnel and of time on mean panel scores for major qualities in hexanethiol (0.025% in dinonyl phthalate). Panel size in brackets.

Quality	Occasion				Mixed panel mean (range)
	1	1	2	3	
	(15)	(8)	(8)	(8)	(15 + 10 + 12)
Sulphurous	2.5	2.4	2.5	2.9	2.6 (0.3)
Putrid	2.3	1.8	1.9	2.0	2.0 (0.6)
Garlic	2.0	1.8	0.6	0.6	1.4 (0.9)
Heavy	1.7	1.5	1.0	1.6	1.2 (0.9)
Sickly	0.9	0.5	1.0	0.5	0.7 (0.7)
Sharp, acid	0.5	0.6	0.9	0.8	0.6 (0.2)
Warm	0.5	0.5	0.3	0.7	0.5 (0.1)
Animal	0.3	0.3	0.6	0.6	0.5 (0.5)
Strength	3.7	3.5	3.4	3.3	3.3 (0.4)
Interval (months)			1	14	

mon to all three panels, although showing more variation than on replication within a short period suggest the stability of the technique.

Sensitivity and discrimination are less easy to demonstrate except on a comparative basis. Results from von Sydow et al. (1970) (Table 4) showed that mass concentrations of 1 and 3 μg/litre of ethyl 2-methyl butyrate in a system based on vacuum-stripped bilberry juice were clearly distinguished at the 8%-significance level on at least one quality (blackberry) despite similar odour strengths. Clapperton (1974) also found differences in beers which were not detected by triangle tests which are sensitive tests of difference. Von Sydow & Karlsson (1971) showed that one pair of samples of blackcurrant juice which were not significantly different by triangle test did differ significantly on three of the odour qualities. Thus the technique is at least as sensitive as traditional difference tests whilst providing much more specific and useful information.

Arbitrary scales and reference standard

Improvements in the technique have been made by von Sydow & Karlsson (1971) and Persson et al. (1973). The first consists of using a 10 or 11 point intensity scale which probably gave greater discrimination and more importantly, probably improved the parametric properties of the intensity data over the 6 point scale. The other improvement was the introduction of a reference standard plus mean values of personal previous evaluations of that reference by the individual assessor. Although the results (Persson et al., 1973) show an increase in variance of most qualities, more qualities were significantly distinguished as different in the two meat preparations when the reference was used than when no reference was used. Thus, although the addition of a reference increased the amount of assessment and therefore probably increased variance the net gain seemed advantageous.

Possible improvements in the method

In the future the method may well be improved in three main ways. Some of the variance is probably caused by deviations in the data from parametricity. Nonparametric methods of analysis, although generally more complex than parametric methods may well aid analysis, particularly of data describing characters containing trigeminal features. Åkesson (1972) has made some studies of analysis of data.

The second and more fundamental improvement is likely to come from further studies of odour classification. Many qualities needed by assessors are correlated to different degrees. Ideally, the qualities used should be independent and further knowledge of the real primary and secondary qualities of odour is likely to reduce the extent to which the technique is dependent upon 'object-language' and overlapping or interdependent qualities.

The third improvement will be the greater utilization of all the data. At present panel means are used and much of the information arising from *differences* between people is not utilized, although an extension of multivariate methods as used by Palmer (1974) or Brown et al. (1975) would allow this. Harries (1973) discussing such methods suggested: 'perhaps we should try, in sensory work, to accept that high variance is synonymous with a high level of sensory information'.

Finally, to what extent is the technique capable of being replaced by instrumental methods? At present, its use to determine significant chemical constituents will allow some degree of instrumental analysis for prediction of sensory properties. So much work using sensory methods is required before this stage is reached, however, that unless great improvements are made it is unlikely that the sensory methods will be replaced except in a few special circumstances.

Acknowledgments

Data in Tables 1, 2 and 5 were taken from a collaborative study on odour description and classification by R. Harper, E. C. Bate-Smith, D. G. Land and N. M. Griffiths.

References

Åkesson, C. A., 1972. Studies in ratio estimation. Ph. D. thesis, Psychological Labs Univ. of Stockholm.

Amoore, J. E. & D. Venstrom, 1966. Sensory analysis of odour qualities in terms of the stereochemical theory. J. Food Sci. 31: 118–128.

Brown, D. G. W., J. F. Clapperton & C. E. Dalgleish, 1975. The language of flavour and its use in product specification. Proc. Amer. Soc. Brew. Chem. 1: 4.

Caul, J. F., 1957. The profile method of flavour analysis. Adv. Food Res. 7: 1–40.

Clapperton, J. F., 1973. Derivation of a profile method for sensory analysis of beer flavour. J. Inst. Brew. 79: 495–508.

Clapperton, J. F., 1974. Profile analysis and flavour discrimination. J. Inst. Brew. 80: 164–173.

Crocker, E. C. & L. F. Henderson, 1927. Analysis and classification of odours. Am. Perfum. essent. Oil Rev. 22: 325–327, 356.

Harper, R., E. C. Bate-Smith, D. G. Land & N. M. Griffiths, 1968a. A glossary of odour stimuli and their qualities. Perfum. essent. Oil Rec. 59: 22–37.

Harper, R., D. G. Land, N. M. Griffiths and E. C. Bate-Smith, 1968b. Odour qualities: a glossary of usage. Brit. J. Psychol. 59: 231–252.

Harries, J. M., 1973. Complex sensory assessment. J. Sci. Food Agric. 24: 1571–1581.
Jones, F. N., 1968. In: N. Tanyolac (Ed.), Theories of odour and odour measurement. Tanyolac (Publ.) Robert College. Istanbul. p. 133–141.
Land, D. G., R. Harper & N. M. Griffiths, 1970. An evaluation of the odour qualities of some stimuli proposed as standards for odour research. Flavour Ind. 1: 842–846.
McCrédy, J. M., J. C. Sonnemann & S. J. Lehmann, 1974. Sensory profiling of beer by a modified QDA method. Food Technol. 28(11): 36–41.
Palmer, D. H., 1974. Multivariate analysis of flavour terms used by experts and non-experts for describing teas. J. Sci. Food Agric. 25: 153–164.
Paukner, E., 1965. Success and failure of odour classification as applied to reactions to erogenous odours. J. Soc. Cosmet. Chem. 16: 515–526.
Persson, T., E. von Sydow & C. Åkesson, 1973. Aroma of canned beef: sensory properties. J. Food Sci. 38: 386–392.
Pilgrim, F. J. & H. G. Schutz, 1957. Measurement of the qualitative and quantitative attributes of flavour. Nat. Acad. Sci. Nat. Res. Counc. Symp. Chemistry of natural food flavours, p. 47–58.
Schutz, H. G., 1964. A matching standards method for characterizing odour qualities. Ann. N. Y. Acad. Sci. 116: 517–526.
Stone, H., J. Sidel, S. Oliver, A. Woolsey & R. C. Singleton, 1974. Sensory evaluation by quantitative descriptive analysis. Food Technol. 28 (11): 24–34.
Sydow, E. von, J. Andersson, K. Anjou, G. Karlsson, D. G. Land & N. M. Griffiths, 1970. The aroma of bilberries (*Vaccinium myrtillus* L.) 2. Evaluation of the press juice by sensory methods and by gas chromatography and by mass spectrometry. LebensmWiss. Technol. 3: 11–17.
Sydow, E. von & G. Karlsson, 1971. The aroma of blackcurrants. 5. The influence of heat measured by odour quality assessment techniques. LebensmWiss. Technol. 4: 152–157.
Yoshida, M., 1964. Studies of psychometric classification of odours. 5. Jap. Psychol. Res. 6: 145–154.

Proc. int. Symp. Aroma Research, Zeist, 1975. Pudoc, Wageningen.

Thiamine, thiamine diphosphate and 'aroma values'

Short communication

J. Solms

Swiss Federal Institute of Technology, Department of Food Science, Universitätstr. 2, CH-8006 Zürich, Switzerland

Thiamine and thiamine diphosphate are important constituents of *Saccharomyces* yeast. They both have a bitter and sulphurous taste, often called 'yeasty taste'. Table 1 (a, b) presents taste thresholds of pure thiamine (T) and thiamine diphosphate (TDP), average mass fractions of both compounds in yeast, and quotients of mass fractions in yeast to that of the pure substances at the taste threshold, called 'aroma values'. The 'aroma values' of pure thiamine and thiamine diphosphate in yeast are relatively small, suggesting a minor contribution to flavour. Table 1 shows also the taste threshold of a yeast preparation (Y), which contains unidentified substances in a mixture (c). In *Saccharomyces*, thiamine and thiamine diphosphate occur together in a proportion 1 : 10 by mass. In such a mixture, there is an appreciable decrease in thresholds and a corresponding increase in 'aroma values' for T and TDP (Table 1, d). If the yeast preparation (Y) is added to the mixture (T + TDP + Y, 1 : 10 : 20), there is a further decrease in the thresholds

Table 1. Sensory evaluation of thiamine and thiamine diphosphate in different systems at pH 5.0. Aroma value is the quotient of mass concentration in yeast to that at the taste threshold.

	Concn at thresholds (mg/kg)	Interaction values (I)[1]	Content in yeast (mg/100g)	Aroma values (A) in yeast
a. Thiamine (T)	0.41		0.2	4.8
b. Thiamine diphosphate (TDP)	8.75		2.0	2.3
c. Yeast prepn (Y)	1.48			
d. T + TDP (1:10)		0.26		
Thiamine	0.047			27
Thiamine diphosphate	0.74			27
e. T + TDP + Y (1:10:20)		0.51		
Thiamine	0.03			67
Thiamine diphosphate	0.3			67
Yeast prepn	0.6			

1. See Appendix 1.

and an increase in 'aroma values' (Table 1, e). This finally results in a 14-fold increase for thiamine and a 29-fold increase for thiamine diphosphate, relative to the pure substances. These effects are due to flavor potentiation, as can be seen from the interaction values (I) which amount to 0.26 for the system T + TDP and to 0.51 for the system T + TDP + Y; the interaction values were calculated on the basis of the pure substances by a method based on Guadagni's method (Appendix 1).

Although thiamine and thiamine diphosphate occur in yeast in concentrations around the thresholds of the pure substances, they probably contribute to the yeasty flavour, by synergism in the system. 'Aroma values' are useful to describe these effects. If combined with 'interaction values', they permit an even better description of the system. More detailed reports have been submitted for publication (Höhn & Solms; Höhn et al., to be published).

References

Guadagni, D. G., J. C. Miers & D. W. Venstrom, 1969. Concentration effect on odour addition or synergism in mixtures of methyl sulfide and tomato juice. J. Food Sci. 34: 630–632.

Höhn, E. & J. Solms. Untersuchungen über Geschmackstoffe der Hefe. 1. Isolierung von Geschmackskomponenten, LebensmWiss. Technol., to be published.

Höhn, E., J. Solms & H. R. Roth. Untersuchungen über Geschmackstoffe der Hefe. 2. Sensorische Beurteilung von Thiamin und Thiamindiphosphat. LebensmWiss. Technol., to be published.

Appendix 1. Calculation of interaction value, based on Guadagni's method (1969).

$$I = (\omega_{B,mix}/\omega_{B,alone}) + (\omega_{C,mix}/\omega_{C,alone})$$

where I is interaction value, ω is mass fraction of components B,C . . . in an aqueous solution at detection threshold either with mixtures of B,C . . . (mix) or alone.

If:

$I < 1$	interaction with potentiating effects
$I = 1$	interaction with additive effects
$1 < I < 2$	interaction with partial additive effects (if no component in the mixture occurs near the threshold concentration)
$I = 2$	no interaction effects
$I > 2$	interaction with antagonistic effects

28 May – Factors governing the emanation of volatile compounds from an odorous substrate

Proc. int. Symp. Aroma Research, Zeist, 1975. Pudoc, Wageningen.

Binding of volatile aroma substances to nutrients and foodstuffs

H. G. Maier

Institut für Lebensmittelchemie der Technischen Universität Braunschweig, D – 33 Braunschweig, Fasanenstrasse 3, Bundesrepublik Deutschland

Abstract

A review is given on the sorption of low-boiling aroma compounds by food components of low water content. Crystallized solids of low molecular weight, e.g. sodium chloride, sugars and amino acids normally have a small sorption capacity. The capacity of glucose and lactose is increased after desiccation of the monohydrates. Aldehydes are bound by several amino acids involving chemical reaction. Products of the Maillardtype are often found in this case. Great amounts of aroma compounds are sorbed by liquid lipids, much smaller amounts by solid lipids. The sorption capacity decreases with the chain length of the fatty acid residues, increases with the number of double bonds. Concerning the proteins the reversible sorption of ethanol, acetone and ethyl acetate depends on the hydrophobicity, the irreversible sorption of ethanol on the polarity of the proteins. Polysaccharides are sorbing in the same order of magnitude as the proteins, but the aroma compounds are sorbed in maximum amount in the presence of water. If polysaccharides were then dried, a part of the aroma compounds is very fast bound by inclusion. Examples for some dry foods are given, especially for coffee extract.

'Binding' is here treated as synonymous with 'sorption' in its broad sense, including adsorption, absorption, and physical and chemical binding.

'Volatile aroma substance' is here any volatile component of a foodstuff with an odour, however weak.

Since foodstuffs are not uniform in structure, it is difficult to study sorption in them. Intact cells bind aroma substances better than broken cells, perhaps because of an outer layer of cutin or an inner layer of suberin. Some volatiles, like methanol, penetrate living cells slower than dead ones. The subject needs further research.

Binding in liquid foodstuffs ought to be simpler and, as often suggested, the vapour pressure of a volatile substance above a liquid foodstuff ought to approach that above the bulk phase; e.g. in the case of aqueous solutions, that above water (Buttery et al., 1971). In preliminary experiments, however, introduction of 0.01 kg food substances per litre of aqueous solutions of volatiles reduced vapour pressure of the volatile by up to 10%; sometimes the reverse occurred (Maier, 1970).

We have not yet investigated why this binding occurs. Especially if foods are moist and if organic structures, such as cells, are present, the problem seems too complex for study as yet. So we have concentrated our attention on dry (or almost dry) foodstuffs and on pure nutrients, for which results are easier to interpret,

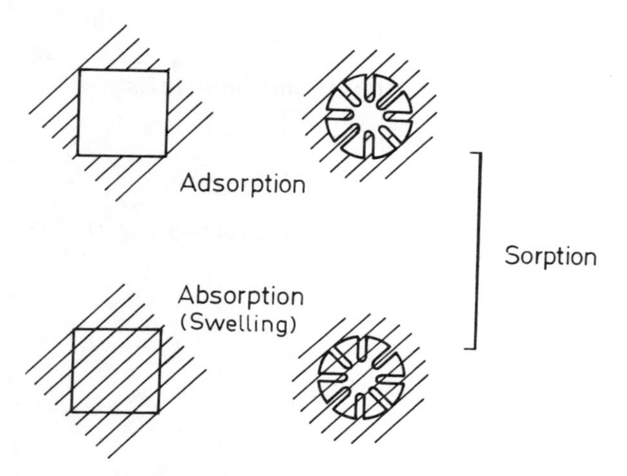

Fig. 1. Types of sorption.

especially as to the nature of binding.

Apart from the effect of biological structures, one would expect that binding of aroma substances in a food consisting of many components would approximate to that of the main ingredients. This proved broadly correct for 5 dried instant foods.

Types of sorption

Dry foodstuffs consist of particles of variable size with an outer surface and, usually, an inner surface made up of fine pores and channels. Volatile substances can therefore be sorbed onto both the outer and the inner surface: this process is called adsorption (Fig. 1). The aroma substance may also penetrate the material of the particle, in effect creating new surfaces: this process is called absorption. If much aroma substance is absorbed, the volume of the particle may increase, the process being swelling.

Table 1. Molality ($mol.kg^{-1}$) of sorbed hexane, acetone or ethanol in pure nutrients exposed to the saturated vapour at 23 °C (desiccator method).

Carrier	*n*-Hexane	Acetone	Ethanol
Triolein	∞	90.29	1.47
Ovalbumin	0.26	0.57	4.99
Potato starch	0.06	0.85	4.52
Lactose	0.35	0.20	1.41
Glucose	0.27	0.22	0.50
Sucrose	0.04	0.02	0.03
Sodium chloride	0.02	0.02	0.02

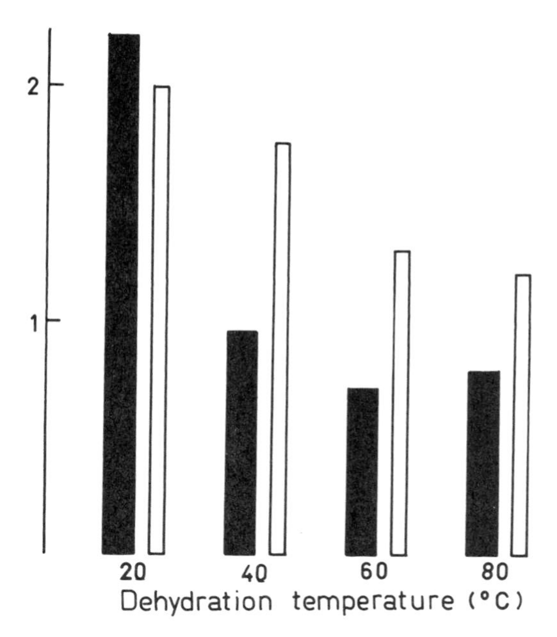

Fig. 2. Black: amount of propionic acid (10^{-2} mol.kg^{-1}) sorbed to glucose dehydrated at different temperatures. White: specific area of surface (10^{-3} m^2.kg^{-1}) sorbing nitrogen.

Pure nutrients

In crystalline nutrients of low molecular mass, the process is mainly adsorption to the outer surface. Pores play no great role. Table 1 reports data from tests on the amounts of three volatiles sorbed to various nutrients.

Triolein, ovalbumin and potato starch are typical of the classes of substances that sorb most: lipids, proteins and polysaccharides. In general, sorption depends on polarity: polar macromolecular substances absorb the polar volatile ethanol most; apolar triolein sorbs hexane most. Acetone falls between ethanol and hexane in polarity and sorption. Crystalline substances of low molecular mass, like sucrose and sodium chloride sorbed the volatiles least of all.

Glucose and lactose seem not to fit into the scheme. This puzzled us until we remembered that they had been prepared by dehydration of monohydrates. Though superficially they looked like normal crystalline sugar, their specific area of surface, estimated by nitrogen absorption, was considerably more than of sucrose and sodium chloride. Further tests showed that specific area depended on temperature during dehydration in vacuo (Fig. 2). The specific area of the monohydrate was only 200 $m^2 \cdot kg^{-1}$ against 1 000–2 000 $m^2 \cdot kg^{-1}$ for the anhydrous form. In electron micrographs (Photo 1), the surface of the monohydrate looked smooth but that of the glucose dehydrated at 20°C showed many small fissures and cleavages, in which one could well imagine extra sorption (Photo 2).

Photo 1. Glucose monohydrate, magnification 8000 ×.

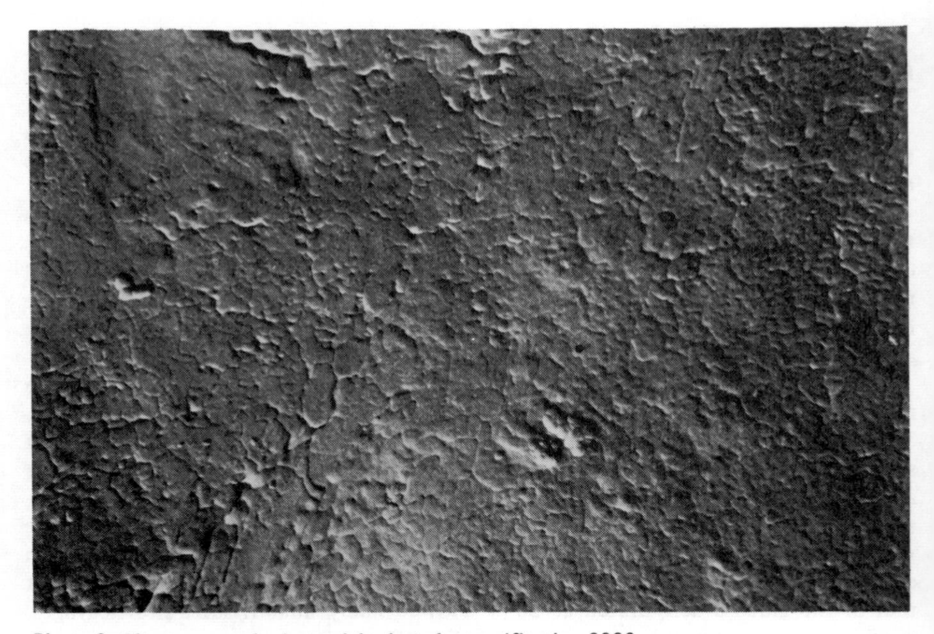

Photo 2. Glucose monohydrate, dehydrated, magnification 8000 ×.

Table 2. Molar heat of sorption (kcal.mol^{-1} = 4.184 kJ.mol^{-1}) estimated by gas chromatography at 30–60 °C.

Volatile	System			
	glucose	lactose	sucrose	sodium chloride
Atmosphere of dry helium				
n-Pentane	5.6	7.0		
Acetone	8.8	8.4	9.8	20.0
Ethyl acetate	9.6	9.5	12.9	11.5
Ethanol	11.3	10.2	12.4	9.2
Atmosphere of moist helium				
n-Pentane		4.9		
Acetone	11.5	13.4	18.8	38.6
Ethyl acetate	10.6	12.2	13.3	20.2
Ethanol		14.0	17.8	
Atmosphere of dry air				
n-Pentane	7.0	8.1		
Acetone	9.6	9.5		16.7
Ethyl acetate	10.1	10.7		17.4
Ethanol				18.4

Binding is normally physical, according to molar heat of sorption at low vapour pressures of the volatiles (Table 2). For butylamine, ethyl acetate and ethanol, molar heat of sorption was close to molar heat of condensation. Only for acetone in dry helium did the value reach 20 kcal·mol^{-1} (ca 85 kJ.mol^{-1}), the limit usually set between physical and chemical binding. If the helium contained a little moisture vapour, the limit was significantly exceeded, as found also for potassium bromide. But in vacuo, the volatiles were easily desorbed from the sugars and sodium chloride, apart from very small residues detectable only with radioactively labelled volatiles. The residue corresponded roughly with 1–2 monolayers at crystal surfaces and could easily be removed with a stream of inert gas at higher temperature.

Amino acids

In special cases, even crystalline substances can bind very large amounts of aroma substances, under some conditions irreversibly. This we observed especially with some amino acids. The formation of thiazolidinecarboxylic acids from cys-

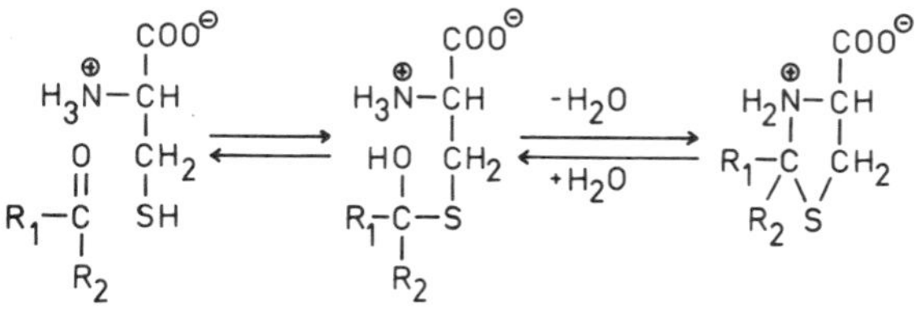

Fig. 3. Reaction between cysteine and carbonyl compounds.

teine is well known in solution (Fig. 3). We demonstrated it also between vapour from volatiles and solid cysteine or cysteine chloride, and also in 'simulated foodstuffs', for instance cellulose powder with a mass fraction of 0.01 cysteine. Thiazolidinecarboxylic acids break down when heated in aqueous solution, especially if acid, so that the reaction has been proposed for flavouring of protein concentrates with aldehydes, which are particularly important aroma substances.

Amounts of volatile carbonyl compounds sorbed by alanine, valine and leucine were mostly intermediate between those sorbed by glycine and isoleucine (Table 3). In general, threonine behaved like serine; methionine like cysteine; aspartic acid and asparagine like glutamic acid and glutamine; lysine like arginine; histidine and tryptophan like phenylalanine. It is noteworthy that amino acids with hydroxyl groups sorbed little. Lysine sorbed most of all.

In these tests, the product often turned brown, perhaps because of Maillard-type reactions. We often observed aldol addition and condensation products too.

Volatile acids and amines were most commonly strongly bound to amino acids. At least in glycine and glutamic acid, which we studied most, other aroma substances were bound in only small amounts, as for sodium chloride.

Table 3. Mass quotient (10^{-3}) of sorbed hexanal, 2-hexenal, acetone and diacetyl exposed to the saturated vapour at 23 °C.

Amino acid	Hexanal	2-Hexenal	Acetone	Diacetyl
Gly	8	12	6	1
Ile	420	30	4	0
Ser	31	0	2	1
CySH	799	535	164	572
Glu	10	0	4	1
$GluNH_2$	11	2	16	1
Arg	1164	17	0	924
Phe	710	75	2	12
Tyr	12	3	6	3
Pro	1252	2038	7	2662
HyPro	5	0	3	1

Table 4. Molality ($mol.kg^{-1}$) of sorbed acetone in lipids at 23 °C (desiccator method).

Coffee, Tanganjika brand	4.0
Coffee extract	0.3
Coffee oil	63.8
Cholesterol	0.9
Trilaurin	1.3
Lecithin (from egg) (Merck)	4.0
Tributyrin	93.5

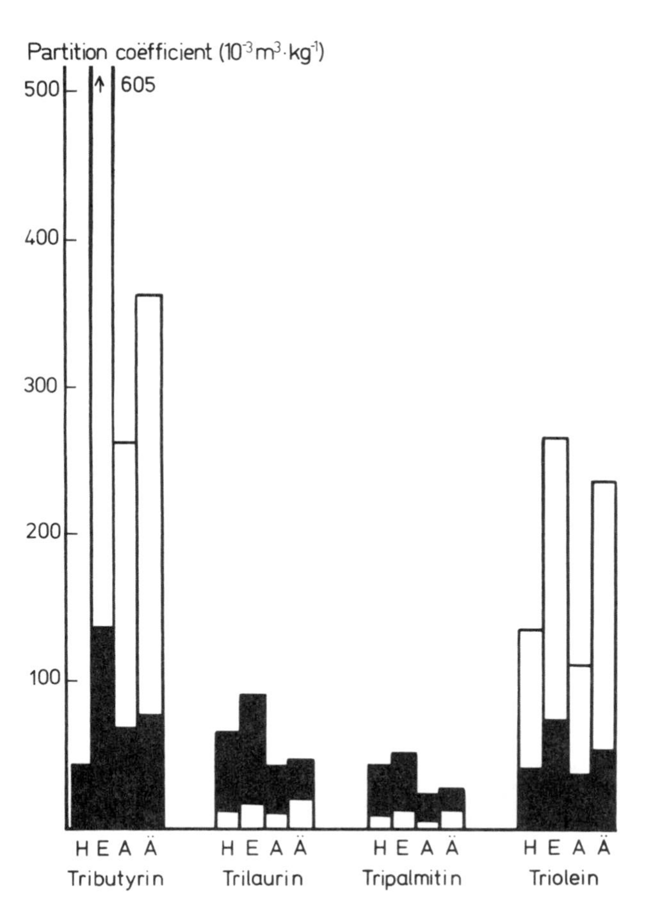

Fig. 4. Partition coefficients of hexane (H), ethyl acetate (E), acetone (A), ethanol (Ä), between triglycerides and helium, at 30 °C (white) and 70 °C (black). Values were calculated from retention times of the aroma substances in gas chromatograms with the lipids as stationary phase.

Triglycerides

Unlike most other substances of low molecular mass, triglycerides sorb rather large amounts of aroma substances, as has long been known. The liquid lipids coffee oil and tributyrin sorb much acetone; the solid lipids cholesterol, trilaurin and lecithin sorb much less (Table 4). Sorption in whole coffee could be chiefly due to coffee oil. Coffee extract, which is almost lipid-free, binds much less.

Partition coefficients too show that solid glycerides sorb less than liquid ones, but they also show that triglycerides with larger fatty acid residues sorb less and that methyl esters of fatty acids with more double bonds sorb more (Figs 4 and 5).

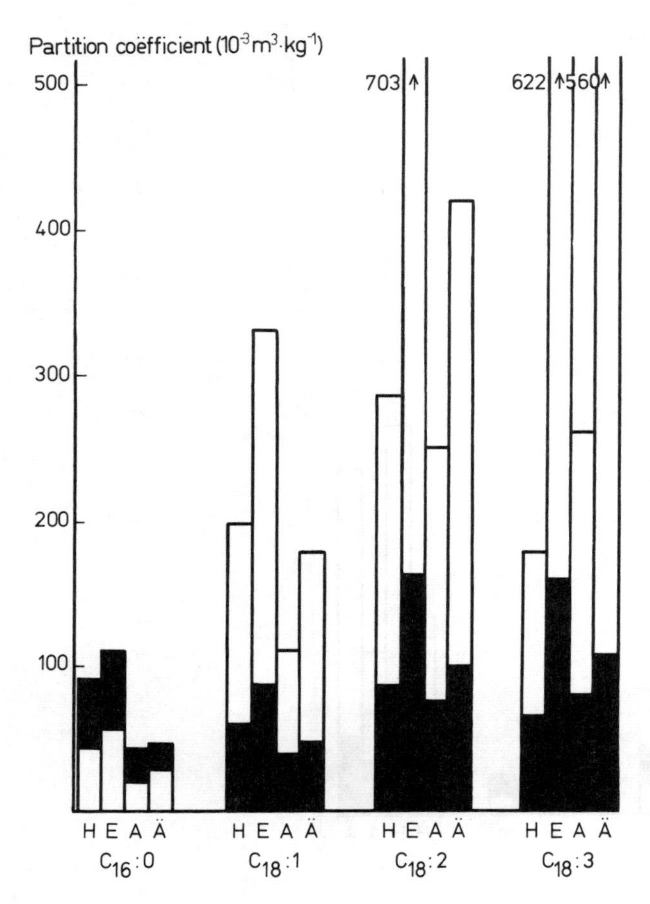

Fig. 5. Partition coefficients of aroma compounds (cf. Fig. 4) between fatty acid methyl esters and helium.

Sorption increases sharply near the melting temperature of any triglyceride, as for tripalmitin (Fig. 6). This property could explain anomalous sorption in biological systems. Trout & McMillan (1943), for instance, found that warm milk sorbed aroma substances more than cold milk, whereas the opposite is true for water and aqueous solutions. This is readily explained by the fact that the fat component in warm milk is liquid.

The shift in peaks of infrared absorption of many systems of lipids and aromatic substances showed that binding is almost always physical. In sorption of phenols, hydrogen bonds are formed.

It seemed plausible that liquid lipids sorb aroma substances in accordance with Fick's law of diffusion, whereas the kinetics of sorption in solid lipids must differ.

Indeed in Fig. 7, the amounts of acetone sorbed by the liquid tributyrin relative to maximum absorption was a rectilinear function of the square root of time; the curve of desorption was also rectilinear. The solid trilaurin finally sorbed much less but most of it was sorbed much quicker, probably on the surface. Later uptake was slower and the process extended over about the same time interval as for tributyrin. We found the same non-Fickian type of diffusion in other solids too.

Proteins

We found a near-rectilinear increase in molality of sorbed component with the square root of time, with ethanol in some proteins (Fig. 8), i.e. a non-lipid substrate. This led us to speculate whether in the case of proteins too, hydrophobic

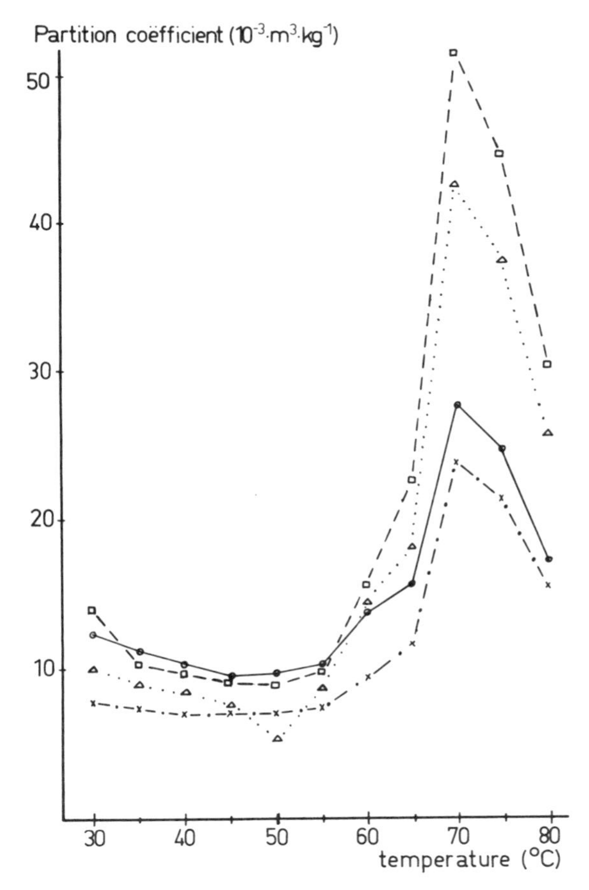

Fig. 6. Partition coefficients of aroma compounds between tripalmitin and helium as function of temperature. x — . — x = acetone, — o — o — = ethanol, △ . . . △ = *n*-hexane, — — □ — — = ethyl acetate.

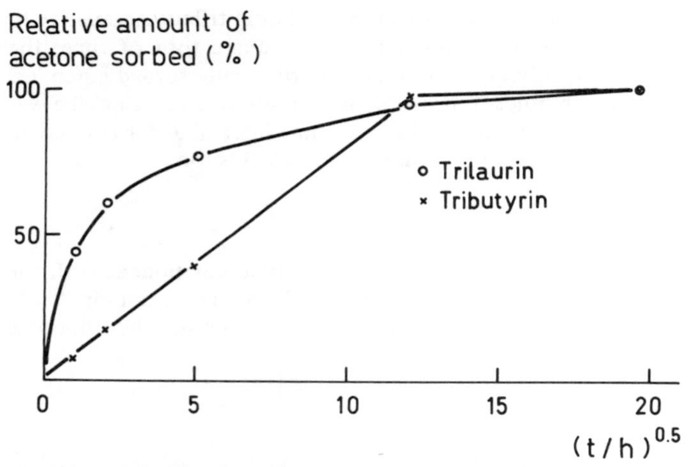

Fig. 7. Amount of acetone sorbed relative to final maximum in triglycerides exposed in a desiccator to saturated acetone vapour as a function of the square root of time in hours.

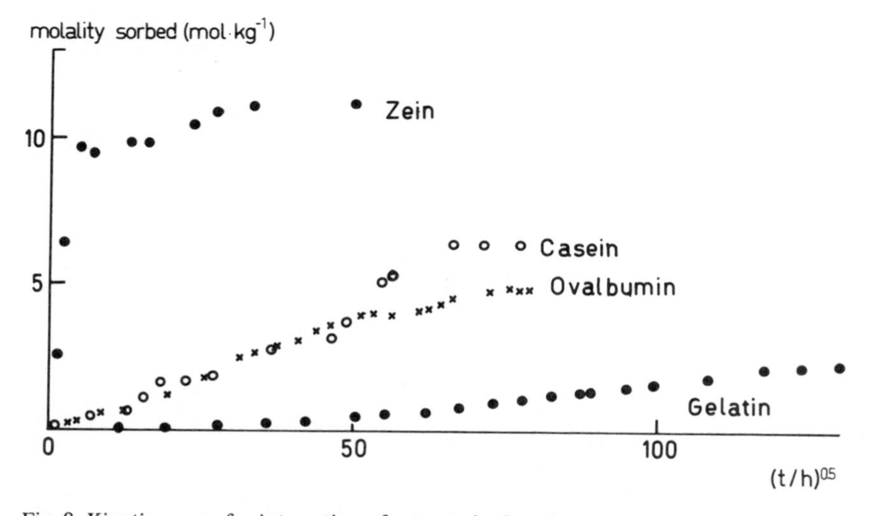

Fig. 8. Kinetic curves for interaction of saturated ethanol vapour with proteins.

interactions might not also play a role. Fig. 9 shows the maximally sorbed amounts of hexane, ethyl acetate, acetone and ethanol, to gelatin, ovalbumin, casein and zein, in desiccator experiments. They increase in this order, in which also increases the hydrophobicity of the proteins. The irreversibly sorbed amounts, indicated by black column in the figure, increase, however, in the opposite order. When one calculates the hydrophoby and polarity, respectively, of the individual proteins, one finds (Figs. 10–11) that, at least with ethanol, and less pronouncedly with acetone

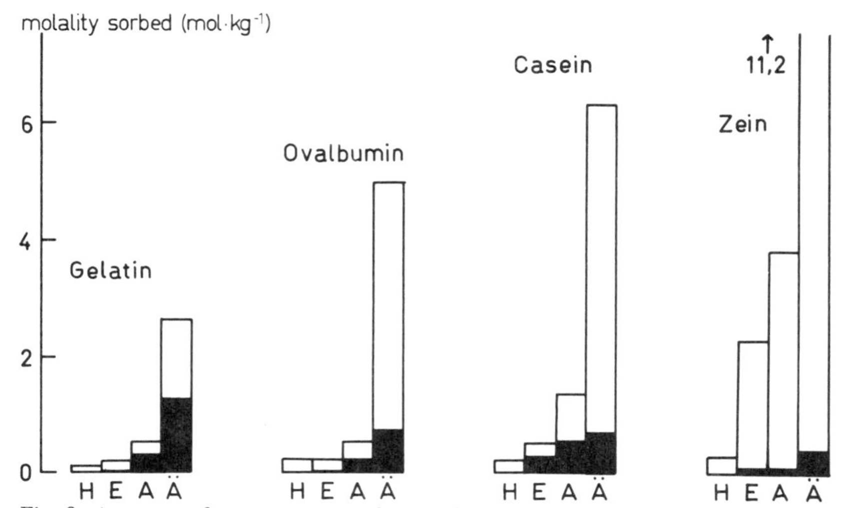

Fig. 9. Amounts of aroma compounds sorbed to proteins, after maximal sorption (white and black) and desorption (black), at 23°C. H = *n*-hexane, E = ethyl acetate, A = acetone, Ä = ethanol.

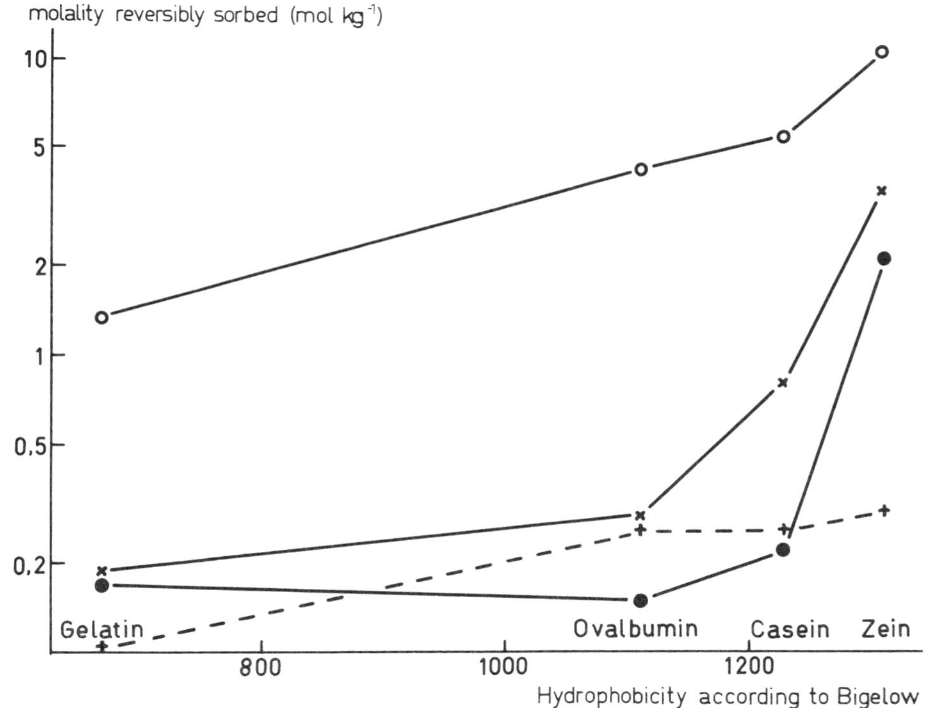

Fig. 10. Reversibly bonded quantities of volatile substances during the application of saturated vapours (23°C) as a function of hydrophobicity of proteins. +---- + = *n*-hexane. ● – ● = ethyl acetate, x — x = acetone, ○ — ○ = ethanol.

and ethyl acetate, there exists a relation to the quantities sorbed. It would seem, therefore, that large amounts of aroma substances are taken up by hydrophobic interactions and are weakly bound, while small amounts can be bound comparatively strongly through polar interactions.

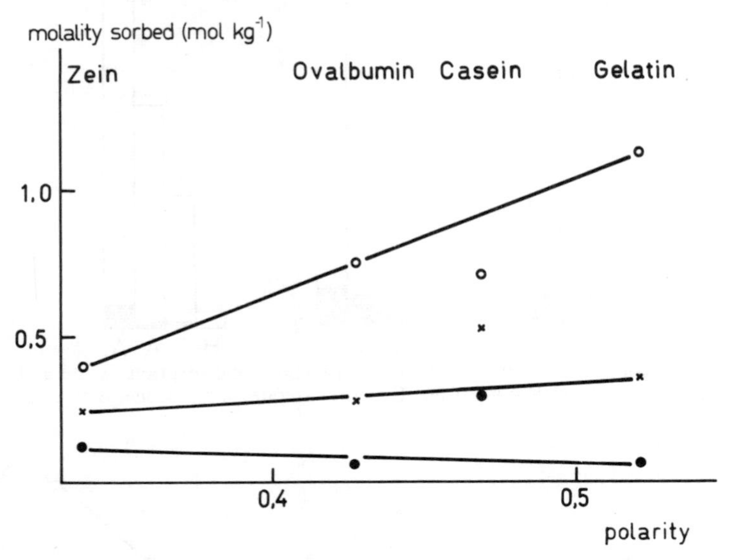

Fig. 11. Irreversibly bonded quantities of volatile substances (23 °C) as a function of polarity of proteins. • — • = ethyl acetate, x — x = acetone, ○ — ○ = ethanol.

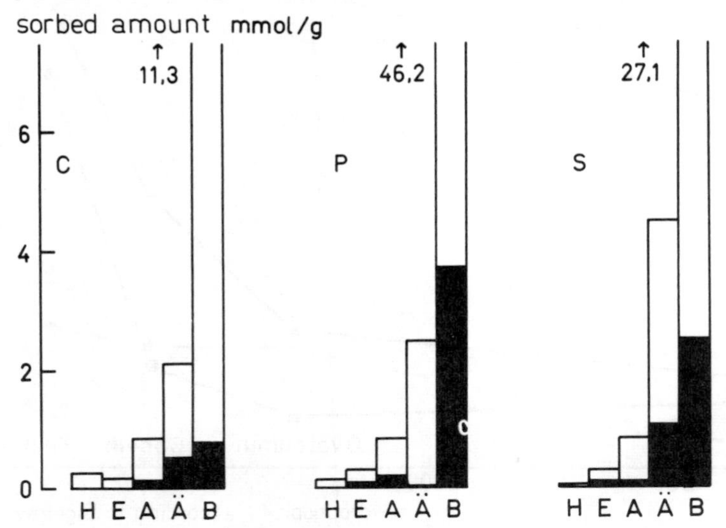

Fig. 12. Aroma substances sorbed to polysaccharides after maximal sorption (white + black) and after desorption (black), at 23 °C. H = *n*-hexane, E = ethyl acetate, A = acetone, Ä = ethanol, B = *n*-butyl amine; C = cellulose; P = pectin; S = starch.

Macro-molecular carbohydrates in anhydrous condition also sorb aroma compounds of varying polarities in the same order of magnitude as do proteins. In Fig. 12 you see the maximally and irreversibly sorbed quantities corresponding to Fig. 9. Generally speaking, the amounts sorbed to cellulose, pectin and starch are similar to those sorbed to gelatin or ovalbumin. In addition, butylamine was investigated. Naturally, it is well bound to pectin and partially remains bound irreversibly as the salt. This also holds for alginic acid. However, the strong binding to cellulose and starch is surprising. We try to account for this by assuming that butylamine is especially apt to disrupt the hydrogen bridges existing between individual carbohydrate molecules, thereby itself forming relatively strong bridges to the hydroxyl groups of these carbohydrates. This has been long known in the case of the sorption of water. Now if one applies the aroma substances to polysaccharides or proteins that have been swollen in water or that contain at least a few per cent of water, then the aroma substances can also penetrate more easily. As an example, in Table 5 you see a few values for the sorption of acetone to dehydrated and to air-equilibrated solids. The relative pressure of acetone vapour amounted to about 0.1% relative. Whereas the differences in sorption are not large in dehydrated circumstances, they are considerable in the air-equilibrated condition. If one now dries such water and aroma compound-containing macro-molecular substances, then especially with polysaccharides, one finds that a part of the aroma substances is very tightly bound. This bonding is stable in vacuum and at elevated temperature, but is anihilated quickly and easily by adding water. In bound condition, no strong interaction between aroma substance and substrate exists. Fig. 13 is an infrared wavenumber spectrum of starch films and of ethyl acetate. After binding of ethyl acetate to starch, only the carbonyl valence stretching band of the ester has shifted to lower frequency, which points to hydrogen bridging to a hydroxyl group of the starch. Otherwise the spectrum is practically unchanged. The aroma compound is thus not destroyed by chemical changes, which is very important for non-destructive bonding purposes. Such inclusion bonds, such as were found for the first time in cellulose by Staudinger (1953), and later in starch by Ulmann & Schierbaum (1958), we were also able to demonstrate in pectin, alginic acid, agar and tylose.

Table 5. 'Partition coefficients' for acetone in sealed-flask experiments. 'Partition coefficient' $k = \dfrac{\text{amount sorbed per gram of solid (g/g)}}{\text{amount of aroma substance in 1 ml of gas phase (g/ml)}}$.

Solid	Dehydrated	Air-equilibrated	Water content, air-equilibrated (%)
Cellulose	51	183	5.2
Dextrin	47	146	9.0
Starch	11	250	8.2
Glucose	10	13	9.0
Lactose monohydrate	4	4	0.3 + 5.0[1]
Pectin	3	159	10.8
Saccharose	2	2	0.0

1. The second term refers to the crystal water.

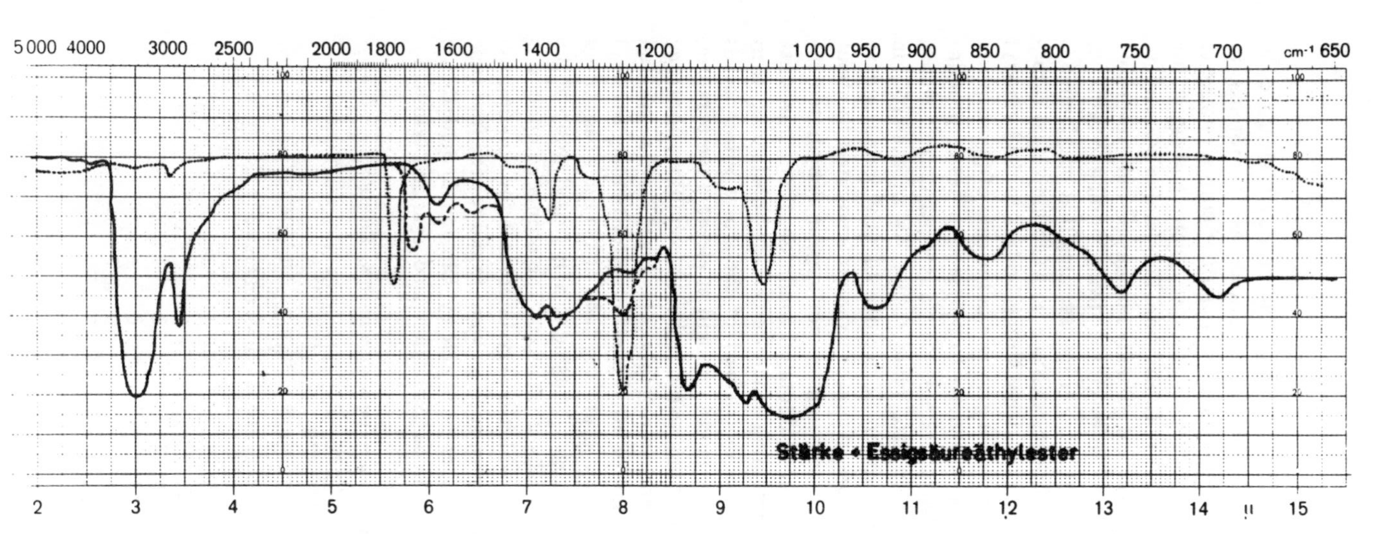

Fig. 13. Infra-red spectra. .... = ethyl acetate, ——— = potato starch, – – – – – – = ethyl acetate occluded in potato starch.

Table 6. Molality ($mol.kg^{-1}$) of sorbed hexane, acetone and ethanol in dry foodstuffs exposed to the saturated vapour at 23°C.

	n-Hexane	Acetone	Ethanol
Whole milk powder	5.24	3.10	4.38
Skim-milk powder	0.27	2.93	4.69
Potato flakes	0.13	0.55	0.65
Powdered coffee extract	0.22	0.26	0.06
Strawberry powder	0.13	0.24	0.41

Binding in dry foodstuffs

Binding like this could also take place during the drying of foodstuffs. We suspect it to occur in coffee extract, of the individual investigated components in which, mannan sorbs best. More closely we investigated five instant dehydrated products as regards their sorption in dehydrated condition (Table 6). Generally, their sorption behaviour shows similarities to that of their main components. Thus, fatty and proteinaceous foodstuffs like whole milk powder (24.3% lipids, 25.4% proteins) and skim-milk powder (0.6% lipids, 31.6% proteins) distinctly sorb the most. They are as a rule followed by potato flakes, which show high contents of polysaccharides (72.7%) and proteins (8%). Coffee extract and strawberry powder showed the lowest contents of lipids and macromolecular substances, and indeed they sorbed only slightly in most cases.

It is not always possible, however, to translate results for pure substances to foodstuffs. E.g. the fat in the whole milk powder, which is for about 60% liquid at 23 °C, ought to sorb 15 mol of acetone per kg if calculation is based on the values found with pure fats. Actually, only 3 mol/kg is sorbed, and allowance for the contribution of the protein is not yet made. Determination of retention volumes of acetone on whole milk powder showed the amount bounded depended on the free fat content. The comparatively low value in Table 6 is therefore most probably to be explained from the fact that the aroma compounds cannot penetrate the fat globule, or do so to a small extent only. Even in the case of ground and roasted coffee are the sorbed quantities as calculated from coffee oil not attained. Here, too, not all of the fat seems to be accessible to the aroma substances. From this, one can see that, as mentioned at the beginning, biological structures may have a distinct effect on the sorptivity of foodstuffs.

References

Buttery, R. G., J. L. Bomben, D. G. Guadagni & L. C. Ling, 1971. J. agric. Food Chem. 19: 1045.
Maier, H. G., 1970. Angew. Chem. int. Ed. Engl. 9: 917.
Trout, G. M. & D. Y. McMillan, 1943. Mich. Agric. Exp. Stn Tech. Bull. 181: 26. Ref. Chem. Abstr. 37 (1943): 4145.
Staudinger, H., K. H. In den Birken & M. Staudinger, 1953. Makromol. Chem. 9: 148.
Ulmann, M. & F. Schierbaum, 1958. Kolloid-Z. 156: 156.

Proc. int. Symp. Aroma Research, Zeist, 1975. Pudoc, Wageningen.

Method for encapsulation of polar compounds in foods

H. G. Peer and B. Hoogstad

Unilever Research Duiven, P. O. Box 7, Zevenaar, the Netherlands

Abstract

One of the problems in the processing of foods is loss or migration of flavour resulting in a poor quality, organoleptically imbalanced-products.

Although the problems are different, specific for each product, we need better control of breakdown of odour and taste compounds, interaction of these with other constituents in the product, volatilization of apolar compounds and leaching, migration of polar material.

Polar compounds in an aqueous medium, in particular during processing can be prevented from migrating or leaching out by encapsulation. Absorption has been investigated thoroughly, but does not preserve flavour. The decrease in leaching could be reduced to no more than a factor 2.

Materials to encapsulate polar flavours must fulfill the following requirements: impermeability for polar material and mechanical stability even at high temperatures; edibility; lack of odour, taste and colour.

Fat almost meets these requirements, except that it is unstable at high temperatures. The simplest way, therefore, would be to encapsulate as water/fat emulsions. In general, however, emulsions that are heat stable also bind, for instance, NaCl, so tightly that this is not released and thus not organoleptically observed.

A successful encapsulation technique starting from fat in which mechanical stability was with, for instance, calcium stearate, m.p. 150°C, which is easily miscible with fat and immobilizes added polar crystals. By gradual improvement of the system, release of flavour was delayed by a factor 100 000. The paper discusses the making of the capsules; physical structure; results of measurement of flavour retention; and some examples where this technique could be used.

Introduction

The more we are concerned with food science and technology, the more we are struck by the highly appreciated sensoric properties of fresh natural products and freshly prepared foods in general.

As far as the flavour is concerned, apart from the aroma composition, a variety of factors in flavour can be distinguished to be responsible for this.

Nature obviously has mechanisms to control the behaviour of polar and apolar flavour compounds, the moment and the way they are formed and are stabilized, as well as the way they are released during consumption.

For manufactured foods, the food industry is confronted with decrease in quality due to loss or migration of flavour during:

a. processing and storage of raw materials,
b. manufacture and storage of foods.

It also faces problems of imparting flavour to bland raw materials such as vegetable oils and proteins, in order to provide food products with acceptable sensory properties.

Apart from flavour losses due to volatilization, leaching and decomposition during processing and storage of raw materials, ingredients may react together. All these problems, of course, are very well known and justify the existence of the flavour industry.

Even though we should have a satisfactory range of flavours, we cannot solve the problems for the food industry. We need better control of stability of odour and taste compounds, interaction of these with other food ingredients, volatilization of apolar material, leaching and migration of polar material, and, last but not least, a better understanding of the effect of texture on sensory properties.

Factors in perception of flavour from manufactured foods

To exploit flavours in manufactured foods, many workers have recently studied physical parameters and interaction of flavours with other ingredients and their effects on perception of flavour.

Interaction of flavour compounds with other components of food

Two types of interaction can be distinguished:
- interference in perception,
- chemical or physical binding of flavour compounds by other food ingredients.

We all are aware of the pioneering work by Maier (1972, 1974), Osman-Ismail & Solms (1973), Solms et al. (1973) and Beyeler & Solms (1974) in this field, which was mainly concerned with the interaction of flavour compounds with polysaccharides (inclusion compounds) and soya protein under conditions comparable to food-processing operations. Their aim was a better understanding of flavour performance in foods rich in starch or protein.

Recently other workers entered into this field. In view of serious losses and changes of the flavour upon addition to soya protein products, Gremli (1974) systematically studied the reversible and irreversible interactions of alcohols, aldehydes and ketones with soya protein due to chemical reactions or physical sorption. Later on Gubler et al. (1974) extended these studies to interaction with other food ingredients such as maltodextrin, lactose and butterfat under simulated processing and storage conditions. Interesting contributions in this field were made also by Pangborn et al. (1974) and Pangborn & Szczesniah (1974) who found specific (suppressing, enhancing, modifying) effects of food hydrocolloids such as sodium alginate on the taste and odour intensities of a range of both polar and apolar flavour compounds in aqueous model systems.

Physical aspects

Texture as such no doubt significantly contributes to the way in which aroma is perceived. The appetizing effect of fresh fruits, the effect of viscosity of a beverage and the succulence of meat and also – in the processed area – the texture of ice cream and margarine are typical illustrations.

In addition, mention has to be made in this context to the effect of cell walls[1]

1. Perhaps the most peculiar situation exists in fresh onion and leek (and partly also in tomato, which derive their strong flavour (mainly) from the fact that the flavour in a natural product is present in 'statu nascendi' and is formed enzymatically upon rupture of cell walls.

in fruits and vegetables fixing aroma components and precursors and thus protecting them against oxidation, reactions etc., whereas in meat connective tissue fibres, mainly consisting of collagen, play a similar role. And, in products like ice cream and margarine, perception of aroma is controlled by partition between oil and water phase.

In this context reference is made to the interesting contribution by McNulty (1974), and already earlier by Mackey (1958), about the effect on taste perception by partition of flavour between oil and water phase in emulsions using a flavour release model.

Method for encapsulation of polar compounds

In order to meet the difficulties on flavour release several methods can be used, one of them being encapsulation. Spray drying is one of the most generally used methods in the food industry. This method, however, does not solve the problem of leaching of, in particular, polar flavour compounds in moist to wet systems. Such compounds equilibrate very quickly between the product and its aqueous surrounding, in particular at high temperatures, a process giving products with no flavour distinction from the environment. To prevent leaching of polar material, barriers impermeable to water and (other) polar material have to be introduced. In this paper we describe a technique to encapsulate polar compounds, on one hand preventing them from leaching, migrating or interacting with other food ingredients during processing and storage, and on the other hand to release them upon consumption.

It should be realized that the desired capsule properties are rather difficult to meet; capsules have to withstand processing conditions (such as sterilization and pasteurization) and storage, but must release the flavour during the final step of food preparation (e.g. frying) or consumption.

The flavour capsules must meet the following specifications:

- mechanical stability: no rupture during processing (temperatures up to 120–150 °C) and storage;
- permeability: impermeable to aromatic components and water;
- composition: permitted edible materials;
- perceptibility: unnoticeable in the food, either to the eye or the palate;
- release of flavour: large enough and weak enough to break, for instance during mastication.

Stability and release seem contradictory. It must be considered, however, that chewing pressures are very high (2–3 MPa). Encapsulated flavours must be heat-stable, for unstable flavours give not only undesired flavour character but also reaction products which can expand the capsules.

Oils and fats meet the specifications.

Emulsions of aqueous solutions of aroma components in oil

In a simple test, water-oil emulsions (w/o), stable to processing temperatures up to 100 °C, did not release their dispersed phase on mastication.

A solution of NaCl concentration 0.3 kg/litre was emulsified in vegetable oil to a heat-stable emulsion consisting of equal volumes of oil and NaCl solution. This emulsion, 5% by mass, was added to a soya-protein dope, 95% before texturizing. In the final product, the mass fraction of NaCl was about 0.5% and examination by microscope showed that the emulsion was still intact. The preparation tasted for less salty than a control sample with the same amount of NaCl added by impregnation. The chewing did not break the emulsion.

Dispersions

Since some emulsions are stable at high temperature, we tested dispersions of polar compounds in oils or fats. However, without an emulsifier, the polar component transferred immediately into the aqueous phase.

Then we tried to develop dispersions stable to sterilization but not to mastication, consisting of an aqueous solution of polar material surrounded by a water-repellant layer. The effectiveness of the capsule was tested with ^{14}C-labelled polar compounds such as monosodium glutamate, glucose and ribose.

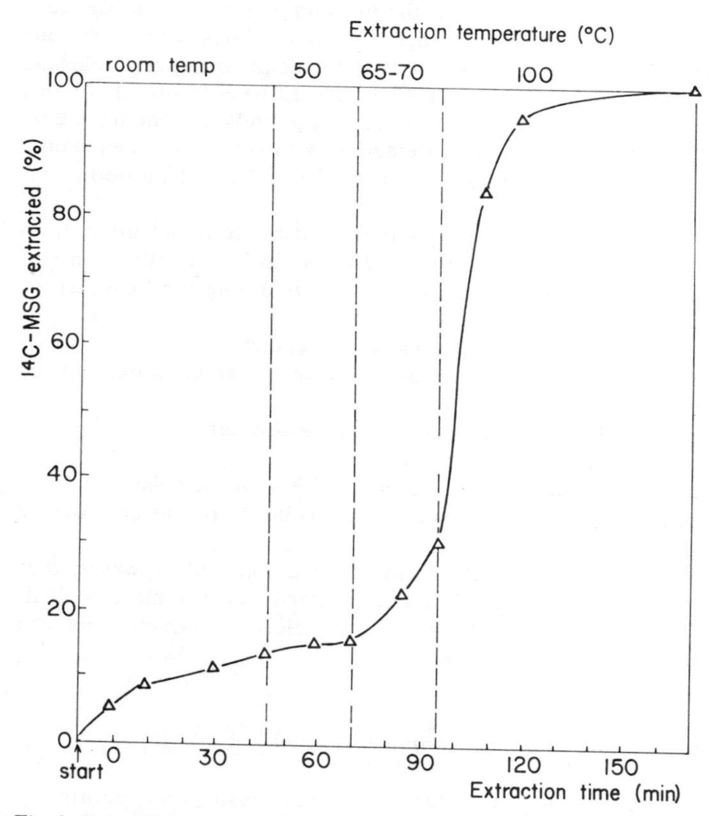

Fig. 1. Extraction at gradually increasing temperatures of [^{14}C]-monosodium glutamate encapsulated in fat flakes (melting point 58 °C).

Capsules of fat of high melting temperature were reasonably effective in cold water. At temperatures above the melting temperature of the fat, the polar component escaped (Fig. 1).

These results show that coatings with higher melting fats would give a good retention also at higher temperatures. Unfortunately, no edible fats or fatty materials are available with melting temperatures above 100 °C, that would prevent the incorporated flavours from leaching at boiling or sterilization temperatures.

Reinforced capsules

The melting temperature can easily be raised by mixing a finely divided solid 'filler' material with a melted fat, to make a thick paste. With oleophilic calcium or magnesium stearate, the 'melting temperature' of the fatty mixture can be raised above 140 °C. Powdered polar flavour compounds mixed with these pastes were well retained in the aqueous phase, in boiling water.

Fig. 2 shows how a particle of fat + filler behaves in water at 100 °C after being mixed with dry flavour particles. Particles made by simply mixing the three components together always lose flavour particles from near the surface to the surrounding water, leaving 'holes' in the mixture, which bring other flavour particles near to the new surface; hence the flavour leaches out slowly. To improve the

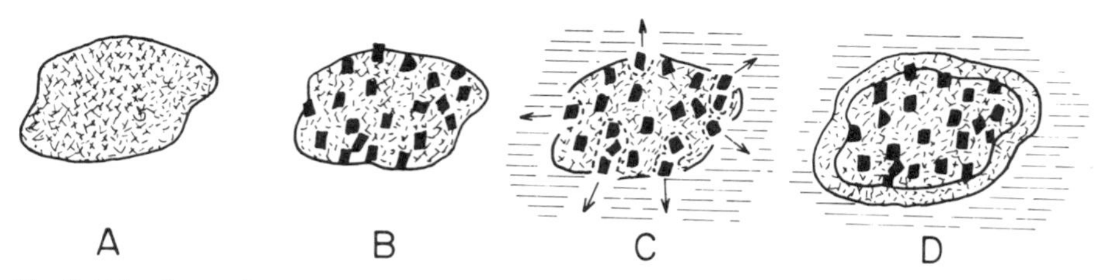

Fig. 2. Behaviour of particles consisting of sunflower seed oil and glucose as filler material. A. Fat + filler; B. Fat + filler + flavour (black particles); C. Particles in water at 100 °C, flavour particles escaping and leaving 'holes'; D. Particles B coated with fat-filler mixture, flavour particles inprisoned.

Table 1. Radiochemical measurement of the time after which 10% (T_{10}) or 50% (T_{50}) of the initial substance fraction of glucose had escaped from a dispersion and from capsules (diam. 4 mm) in boiling water.

Preparation	Ingredients (by mass)	T_{10}	T_{50}
Dispersion	^{14}C-labelled glucose 70% + sunflower seed oil 30%	< 0.01	0.33
Capsule C	^{14}C-labelled glucose 33% + sunflower seed oil 33% + calcium stearate 33%	~ 10	600
Capsule D	As C coated with 0.2 mm calcium stearate and sunflower seed oil	~ 660	≫ 720

Plate. Behaviour of particles 1.5 mm in diameter with sugar (mass fraction 40%), and of fat/calcium stearate (60%) and caramel: a. Coated (light) and uncoated (dark) capsules. b. Cross-section of coated capsules. c. Capsules after 1 h in boiling water. d. Cross-section of boiled capsules.

retention, particles of Type B (Fig. 2) had to be coated with a thin layer of the fat-filler mixture as in Type D.

According to Table 1, retention was 1000 times as good with capsules of Type B, with a mixture of filler and glucose/oil, as with a simple dispersion. A coating gave a further increase of 70 times.

Of course, capsules of diameter 4 mm are of no use for food flavouring. Much smaller capsules can be made by extruding the mixture above the melting temperature of the added fat, producing cylindrical segments (0.5–1.5 mm in diameter). The segments are then cooled, broken into smaller particles and rounded off in a Spheronizer. They are then coated by slowly adding 'filler' as the capsules tumble on the Spheronizer plate. The filler powder sticks to the particle surface and melted fat soaks into it to the desired coat. The size of the spheres is determined by the diameter of the segments from the extruder. The thickness of the coat is controlled by the amount of filler.

The plate shows that coated particles retained the flavour (sugar + caramel) and that uncoated ones lost it, leaving holes.

Coating is obviously essential. Figure 3 shows the favourable effect of proportion of filler on retention of polar compounds. According to Figure 4, spheres lose much less flavouring than cylinders, no doubt because the coating is more even.

To conclude, these capsules can be used for flavouring food. Capsules (diameter 1 mm), containing a water-soluble meat flavour and some salt, were incorporated in soup balls in which part of the meat was replaced by soya protein. These soup balls were boiled for 30 min in water and were assessed organoleptically by a panel of 10 persons. Similar 'meat balls' to which the flavour had been added as

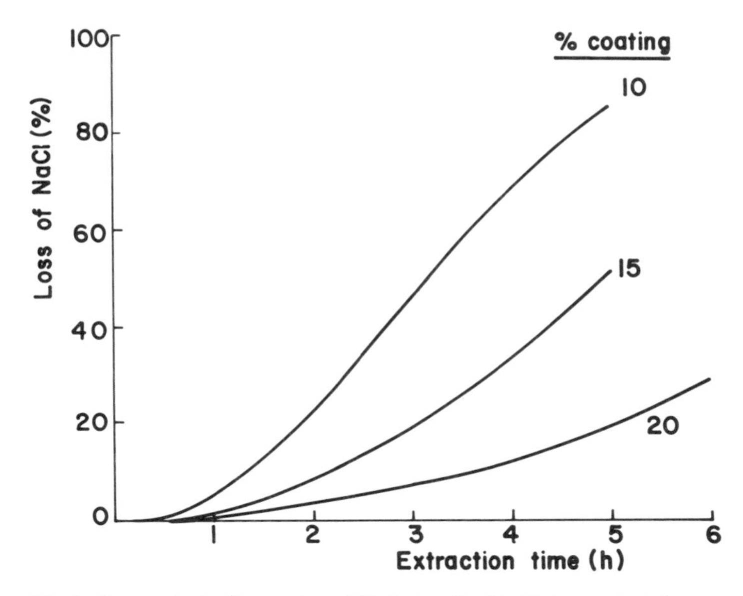

Fig. 3. Escape in boiling water of NaCl (in % of initial quantity) from capsules with different mass fraction of coating (% coating) as a function of time.

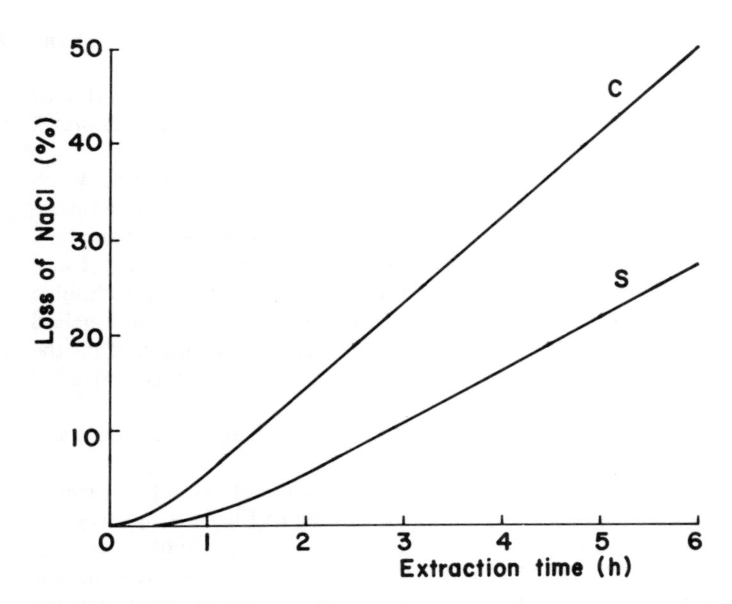

Fig. 4. Escape in boiling water of NaCl (in % of initial quantity) from spherical (S) and cylindrical (C) capsules as a function of time.

such were used as a reference. The whole panel preferred soup balls containing the capsules, which gave a stronger meat-like taste.

References

Beyeler, M. & J. Solms, 1974. Interaction of flavour model compounds with soya protein and bovine serum albumin, Lebensm. Wiss. Technol. 7: 217.

Gubler, B. A., H. A. Gremli, C. Verde & J. Wild, 1974. Some aspects of interactions between food- and flavour-components, 4th Int. Congr. Food Science and Technology, September 1974, Madrid.

Gremli, H. A., 1974. Interaction of flavour compounds with soy protein, J. Am. Oil Chem. Soc. 51: 95A.

Mackey, A. O., 1958. Discernment of taste substances as affected by solvent medium, Food Res. 23: 580.

McNulty, P. B., 1974. Quantitative analysis of flavour threshold data in various media using a flavour release model, 4th Int. Congr. Food Science and Technology, September 1974, Madrid.

Maier, H. G., 1972. Moderne Vorstellungen über die physische und chemische Bindung von Aromastoffen, Dechema Monogr. 70: 323.

Maier, H. G., 1974. Zur Bindung flüchtiger Aromastoffen und Proteine, Deut. Lebensm. Rundsch. 70: 349.

Osman-Ismail F. & J. Solms, 1973. Formation of inclusion compounds of starches with flavour substances, Lebensm. Wiss. Technol. 6: 147.

Pangborn, R. M. & A. S. Szczesniak, 1974. Effects of hydrocolloids and viscosity on flavour- and odour-intensities of aromatic flavour compounds, J. Texture Studies 4: 467.

Pangborn, R. M., I. M. Trabue & A. S. Szczesniak, 1974. Effects of hydrocolloids on oral viscosity and basic taste intensities, J. Texture Studies 4: 224.

Solms, J., F. Osman-Ismail & M. Beyeler, 1973. Interaction of volatiles with food components, Can. Inst. Food Sci. J. Technol. 6: A10.

Proc. int. Symp. Aroma Research, Zeist, 1975. Pudoc, Wageningen.

The effect of process conditions on aroma retention in drying liquid foods

P. J. A. M. Kerkhof and H. A. C. Thijssen

Laboratory for Physical Technology, Department of Chemical Engineering, Eindhoven, University of Technology, P. O. Box 513, Eindhoven, the Netherlands

Abstract

The aroma of foods is imparted by a large number of compounds the majority of which are very volatile. Upon drying part of these volatiles may be lost, thus causing both decrease in aroma strength and a change in the overall impression of the aroma constituents.

In drying processes the removal of water and the loss of aroma are generally controlled by molecular diffusion *inside* the foodstuff. The transport rates are strongly influenced by the structure of the drying specimen. The effect of the structure upon drying rate and more in particular upon aroma losses is discussed for liquid foods. The following structures are distinguished:

- hollow particles (spray-drying)
- solid particles (low temperature air-drying)
- porous particles (freeze-drying of granules)
- solid sheets (air-drying of slabs)
- porous sheets (freeze-drying of slabs).

For these structures the experimentally observed effects of drying conditions upon aroma retention are reviewed for some model solutions of liquid foods. The effects are interpreted by the selective diffusion theory. With this theory the effects of process conditions and the effects of the structure and dimensions of the particles can be explained. For most geometric shapes the aroma retention can even be calculated quantitatively if physical data, including vapour pressure isotherms and relationships between diffusion coefficients and water concentration, are available.

Introduction

One of the primary factors determining the quality of many natural juices and extracts is flavour pattern. This pattern is made up of many aroma components. The vast majority are present at very low concentrations and are very volatile relative to water (Thijssen & Rulkens, 1968; Bomben & Merson, 1969; Thijssen, 1972; Chandrasekaran & King, 1971). Upon equilibrium evaporation of water these components will already be completely removed from the food liquid, when only part of the water has been evaporated (Thijssen & Rulkens, 1968; Thijssen, 1972). However, experiments on the retention of volatile flavour components in several drying processes, including spray-drying, slab drying, freeze-drying and extractive drying, have shown that, under optimum processing conditions, aroma components can be largely retained. Hence for these non-equilibrium processes there are rate-controlling factors limiting aroma loss. Because of the high volatility of the aroma compounds, the main limitations for aroma transport lie inside the drying specimen. This paper briefly discusses the transport mechanisms presented in the literature; it will be seen that both molecular transport properties and structure influence the transport of water and aroma components during dehydration.

Mechanisms for aroma transport

Mechanisms explaining the retention of homogeneously dissolved aroma components are:

1. adsorption of volatiles
2. micro region entrapment of aroma components
3. difference in diffusivity between aroma components and water.

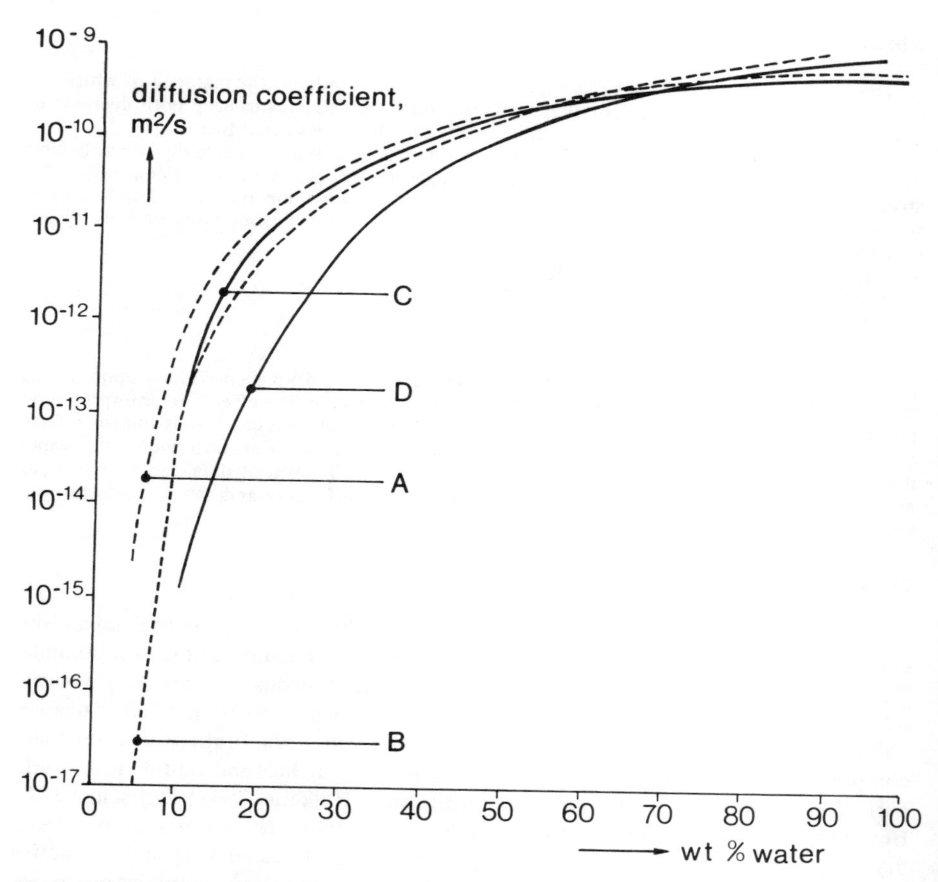

Fig. 1. Effect of mass fraction of water (wt % water) on the diffusion coefficient of water (A) and of acetone (B) in coffee extract (Thijssen & Rulkens, 1968) and of water (C) and of acetone (D) in maltodextrin (Menting, 1969) at 25 °C.

The concept of *adsorption,* as given by Rey & Bastien (1962) for freeze-drying, states that during drying aroma molecules escape from the moist region, and are adsorbed at active sites on the dry interface. Their experimental data and those of other authors (Flink & Karel, 1970a; Issenberg, 1968; Menting, 1969; Capella et al., 1974) show that surface adsorption does not contribute much to aroma retention, but that some of the phenomena found are more likely to be caused by absorption in the (partially) dry material.

The *microregion* concept as formulated by Flink (1969) and Flink & Karel (1969) states that during freezing and freeze drying microstructures are formed by carbohydrate or protein molecules, which are connected by hydrogen bonds. Aroma molecules can be entrapped in these microregions. Many experimental results (Flink, 1969; Flink & Karel, 1969; Karel & Flink, 1973; Flink & Karel, 1970b; Bartholomai et al., 1974, 1975; Chirife & Karel, 1973a, 1973b, 1974; Chirife et al., 1973; Flink & Karel, 1972; Kayaert et al., 1974, 1974a), can be explained qualitatively by this theory. A critical review of these results by King & Massaldi (1974), however, showed that the experimental results can also be explained by molecular diffusion. As they stated, the microregion theory presents a microscopic view of the phenomena which are macroscopically described by the diffusion theory.

The *selective diffusion theory,* as stated by Thijssen (1965), considers the molecular diffusion of water and aroma components inside the drying specimen. In solutions of carbohydrates, gums or proteins, the diffusion coefficients of both water and aroma components depend strongly upon water content. This is illustrated by Figure 1 for aqueous maltodextrin solutions and coffee extract, to which small amounts of acetone had been added as a model aroma component. Analogous results were found by Chandrasekaran & King (1969, 1972). The decrease in the diffusion coefficient of the aroma component with decreasing water content is much sharper than that of water, at low water contents leading to a diffusion coefficient of aroma component several orders of magnitude less than of water. As water is withdrawn at the interface of a drying specimen during dehydration, water concentration gradients build up, and the interfacial water concentration decreases with time. After some time, the interfacial concentration of water becomes so low that the diffusion coefficient of aroma is much lower than that of water, and the interface behaves like a semipermeable membrane, through which only water can diffuse.

Mathematical evaluation of this theory has shown quantitative agreement with experimental data on drying for systems in which aroma is homogeneously dissolved. In liquid foods, the transport and retention of aroma components may also be governed by the presence of one or more dispersed phases, which may be volatile or nonvolatile. Recent developments were reviewed by King & Massaldi (1974) and by Kayaert (1974b). They dealt with both the retention of volatile droplets and of volatile dissolved in an inert disperse liquid phase.

This paper describes the transport of homogeneously dissolved aroma components in different structures during dehydration, and finally some remarks with respect to the effect of a disperse phase are made. The drying of a solid slab will be treated extensively; the behaviour of other types of structure is in many ways analogous.

Isothermal drying of solid slabs

Theory of water transport

A diagram of a solid slab is given in Figure 2(a). Let drying take place from one side. The flux of water in such a system can be described by a binary diffusion equation for water (Menting, 1969; Thijssen & Rulkens, 1968; Chandrasekaran & King, 1969, 1972b):

$$J_w = -D_{ww}\,(\partial \rho_w / \partial r) \tag{1}$$

in which J_w = mass flux of water relative to stationary coordinates
ρ_w = mass concentration of water
r = distance coordinate.

The rate of evaporation at the surface, $J_{w,i}$, is higher than the local flux J_w with respect to stationary coordinates. This is due to the shrinkage due to the water loss:

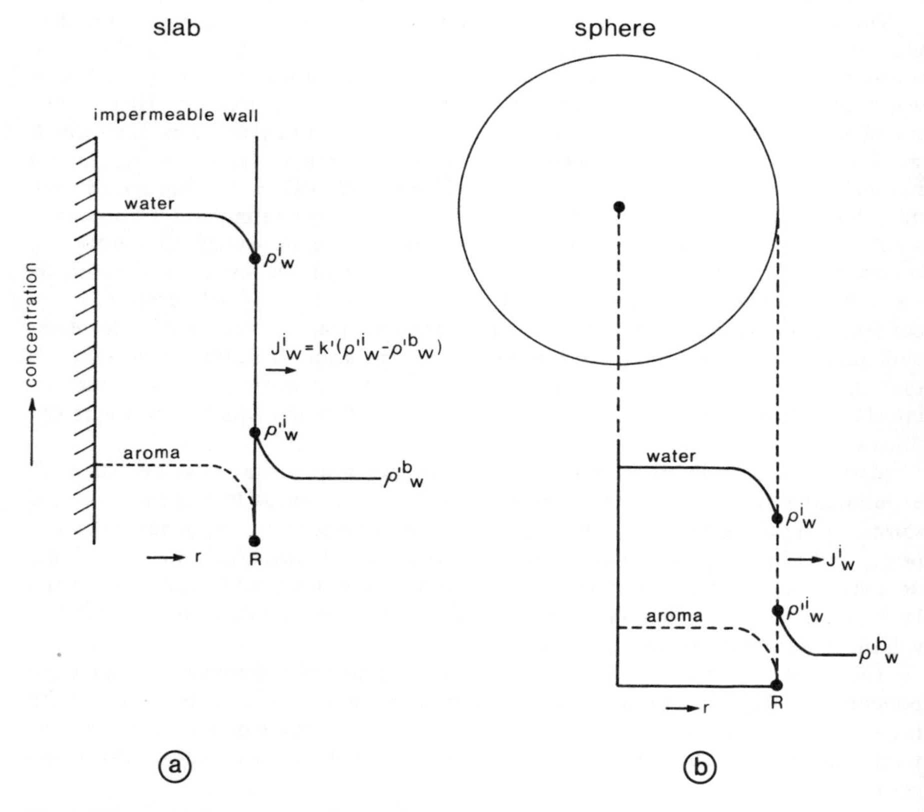

Fig. 2. Diagrammatic representation of drying specimen: (a) slab, (b) solid sphere.

$$J_{W,i} = - D_{WW} (\partial \rho_W / \partial r) / (1 - \rho_W \nu_W) \qquad (2)$$

in which $J_{W,i}$ = flux of water with respect to interface
ν_W = partial specific volume of water.
The differential equation for the water transport reads:

$$\partial \rho_W / \partial t = \frac{\partial}{\partial r} (-J_W) \qquad (3)$$

with initial and boundary conditions:

$$t = 0 \qquad 0 < r < R_0 \qquad \rho_W = \rho_{W,0} \qquad (4)$$

$$t > 0 \qquad r = 0 \qquad \partial \rho_W / \partial r = 0 \qquad (5)$$

$$r = R \qquad J_{W,i} = k' (\rho'_{W,i} - \rho'_{W,b}) \qquad (6)$$

in which k' = mass transfer coefficient in the gas phase

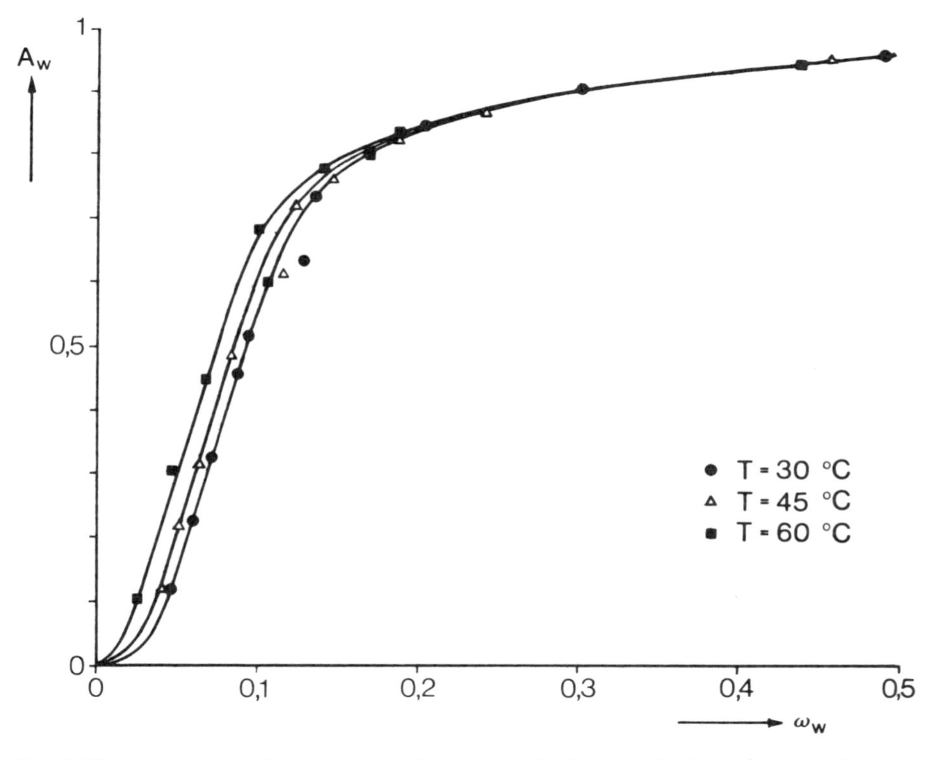

Fig. 3. Water vapour sorption isotherms of aqueous maltodextrin solution at 3 temperatures, as determined by the desiccator method (Kerkhof, 1975). ω_W = mass fraction of water; A_W = water activity.

$\rho'_{w,i}$ = mass concentration of water in the gas phase at the interface
$\rho'_{w,b}$ = bulk mass concentration of water in the gas phase

To solve this set of equations we must know the relation between D_{ww} and ρ_w at the given temperature, and the relation between $\rho'_{w,i}$ and $\rho_{w,i}$, the interfacial mass concentration of water in the liquid phase. This latter relation is given by:

$$\rho'_{w,i} = A_{w,i}\, \rho'_w{}^* \tag{7}$$

in which $A_{w,i}$ is the water activity at equilibrium with $\rho_{w,i}$ and $\rho'_w{}^*$ is the equilibrium water vapour concentration in gas phase with pure water. The relation between $A_{w,i}$ and $\rho_{w,i}$ is given by the sorption isotherm. A typical example is given in Fig. 3 for the system water-maltodextrin (Kerkhof, 1975). The numerical solution of these equations is discussed extensively by the above mentioned authors and by Kerkhof et al. (1972).

Theory of aroma transport

As shown by Chandrasekaran & King (1969, 1972b), Rulkens & Thijssen (1969), Kerkhof et al. (1972), and Rulkens (1973), the transport of each aroma component can be described by a ternary diffusion equation:

$$J_a = -D_{aa}\,(\partial\rho_a/\partial r) - D_{aw}\,(\partial\rho_w/\partial r) \tag{8}$$

in which D_{aa} and D_{aw} are ternary diffusion coefficients depending upon concentration and temperature. Both D_{aa} and D_{aw} decrease with increasing molecular weight of the aroma components. These dependences include relations between activity coefficients and concentrations. Thijssen (1972) showed that for the systems under consideration certain simplifications of the rather complicated relations can be made resulting in:

$$J_{a,s} = -D_{aa}\,\{(\partial\rho_a/\partial r) + \rho_a\,(\partial \ln H_a/\partial\rho_w)\,(\partial\rho_w/\partial r)\} \tag{9}$$

in which $J_{a,s}$ = aroma flux with respect to dissolved solids
H_a = activity coefficient of aroma component on mass basis and defined by $A_a = H_a\rho_a$.

The activity coefficient of aroma components decreases with increasing water concentration (Chandrasekaran & King, 1969, 1970; Thijssen, 1972; Bomben & Merson, 1969). If mass concentration gradients of water are small or H_a is weakly dependent on ρ_w:

$$J_{a,s} = -D_{aa}\,(\partial\rho_a/\partial r) \tag{10}$$

The diffusion equation for aroma transport reads:

$$\partial\rho_a/\partial t = \frac{\partial}{\partial r}(-J_a) \tag{11}$$

with initial and boundary conditions:

$$t = 0 \qquad 0 < r < R_0 \qquad \rho_a = \rho_{a,o} \tag{12}$$

$$t > 0 \qquad r = 0 \qquad \partial \rho_a / \partial r = 0 \tag{13}$$

$$r = R \qquad \rho_a = 0 \tag{14}$$

Boundary condition (14) may be applied because of the extreme volatility of aroma components. As the diffusion coefficients are functions of the mass concentration

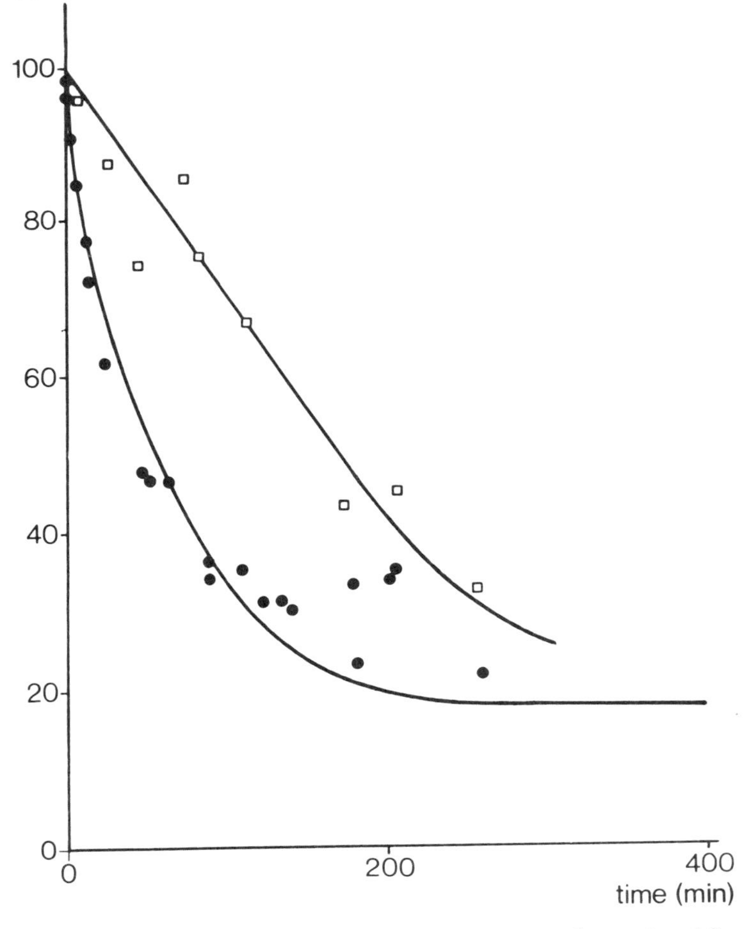

Fig. 4. Comparison of theoretically predicted (lines) and experimentally observed (points) drying rate and aroma loss from a drying slab of aqueous maltodextrin solution (Menting, 1969; Menting et al., 1970a). Vertical axis: mass fraction of acetone (●) and moisture (□).

of water, the solution of the diffusion equation for aroma should be performed simultaneously with the solution of the water diffusion equation. Numerical solutions are discussed by the authors mentioned in the previous section (Theory of water transport).

Comparison of theoretical and experimental results

Menting (1969); Menting et al. (1970a, 1970b), experimentally determined the relation between the diffusion coefficients of water and of acetone in aqueous maltodextrin solution, and performed isothermal drying experiments on the same system. In Figure 4 the points indicate results of his experiments. The solid lines represent data calculated with the experimentally observed diffusion coefficient and the water vapour sorption isotherm. This figure shows that the diffusion model gives an accurate description of the drying process. Rulkens & Thijssen (1969) obtained similar results. Chandrasekaran & King (1969, 1972a, 1972b) experimentally determined ternary diffusion coefficients of water in aqueous solutions, and determined concentration gradients of water and of aroma inside a drying slab. Figure 5 is an example of their calculated and experimentally determined concentration profiles; again good agreement between model and experimental result can be concluded. The maximum in the aroma concentration profile has been discussed by Chandrasekaran & King, and by Kerkhof et al. (1972), who explained it by the influence of the water concentration on the activity coefficient of aroma and by shrinkage of the slab.

Figure 4 clearly shows a period of constant drying rate, and that aroma loss in good approximation takes place during this period. Figure 6 shows the numerically calculated values of the interfacial mass concentration in the liquid phase ρ_w^i, of the interfacial water activity A_w^i and of the aroma retention for a typical slab drying situation (Kerkhof, 1974). While the interfacial concentration decreases continuously the interfacial water activity first remains constant and then starts to decrease rather sharply. This phenomenon can easily be understood from the sorption isotherm. Figure 3 indicates that for mass fractions of water higher than 0.30 the equilibrium water activity exceeds 0.9. Let the value at which the water activity reaches 0.9 be denoted as 'critical water concentration'. Apparently upon passing this critical concentration at the interface at about the same time the interface becomes almost impermeable for aroma transport, as can also be seen from Figure 6. Using this observation, Thijssen & Rulkens (1968) and Menting (1969) calculated the aroma retention under different processing conditions from the length of the constant-rate period and an effective aroma diffusion coefficient $D_{a,eff}$ during this period. The aroma retention can be written as:

$$AR = (8/\pi^2) \sum_{n=0}^{\infty} \{ 1/(2n+1)^2 . \exp -(2n+1)^2 \pi^2 Fo_c/4 \} \quad (15)$$

where

$$Fo_c = (D_{a,eff}\, t_c)/R_0^2 \quad (16)$$

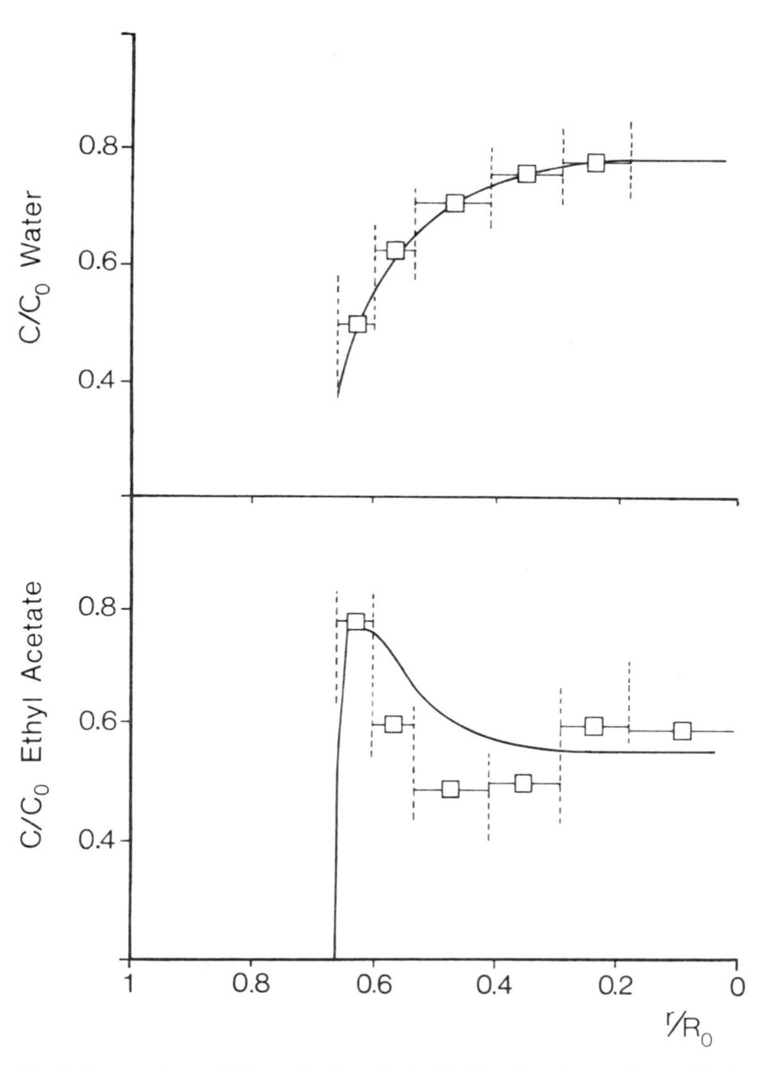

Fig. 5. Comparison of theoretically calculated (lines) and experimentally determined (points) water and aroma profiles in drying aqueous sucrose solution with ethyl acetate as model aroma component (Chandrasekaran, 1969; Chandrasekaran & King, 1972b). r = distance from unpermeable side; R_0 = initial thickness of the slab; c = concentration at time t; c_0 = initial concentration.

and where AR is the quotient of amount of an aroma component left in the drying specimen after drying time $> t_c$ relative to the amount initially present.

This concept was later explored rigorously by Kerkhof (1974, 1975) who correlated the length of the constant-rate period t_c and $D_{a,eff}$ to process variables, and presented a simple prediction method for aroma retention based on this correlation.

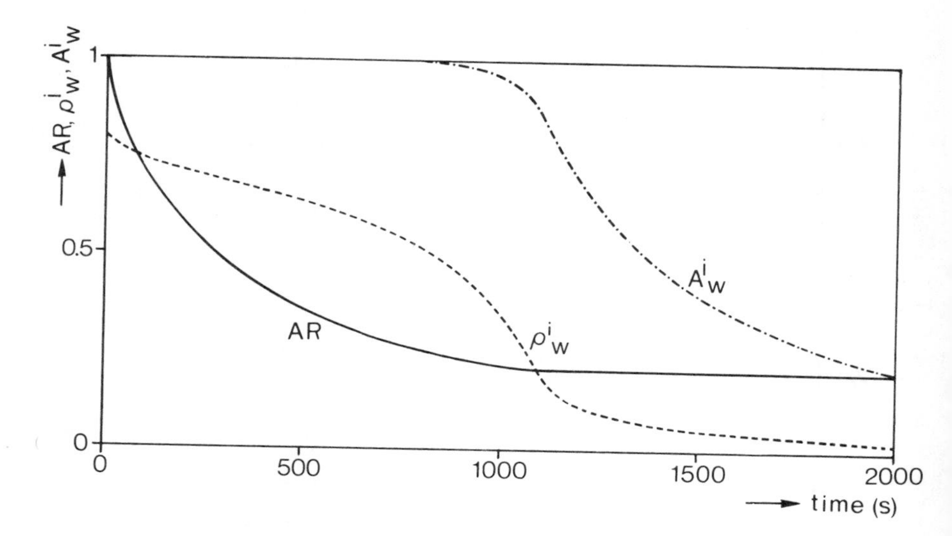

Fig. 6. Theoretically calculated interfacial water concentration ρ^{i}_{w} 10^{-3} kg/m³ and activity $A_{w,i}$, and aroma retention AR in relation to time for slab drying.

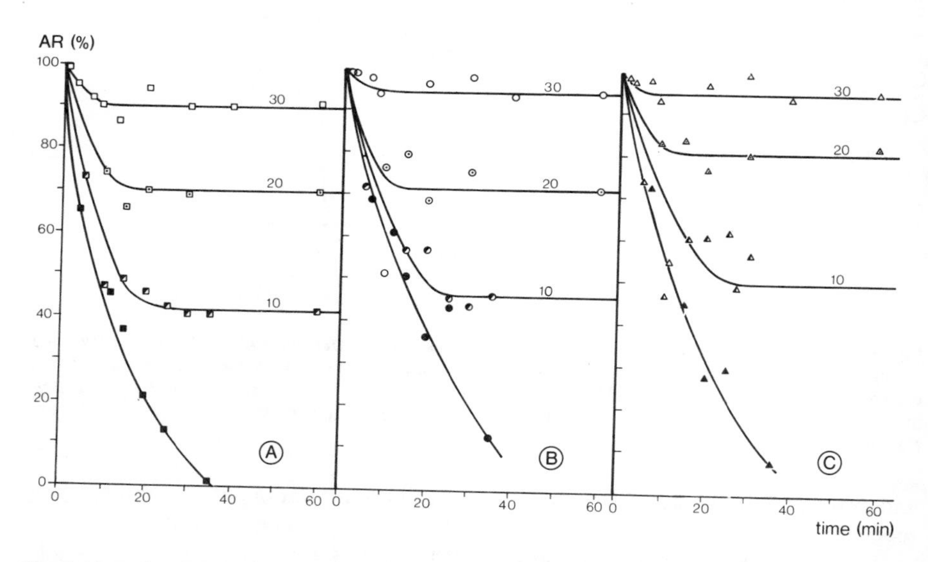

Fig. 7. Experimentally observed retention of *n*-alcohols in drying slabs of aqueous maltodextrin solutions at 30 °C. Parameter initial dissolved solids content (wt %) (Rulkens, 1973). A: methanol; B: *n*-propanol; C: *n*-pentanol.

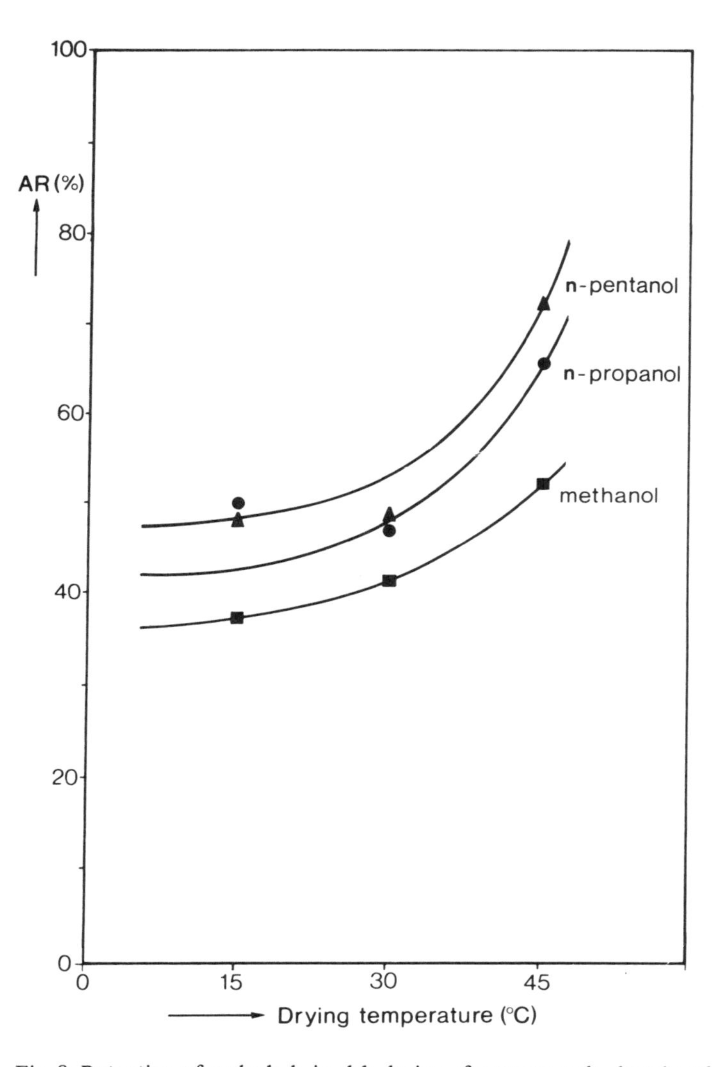

Fig. 8. Retention of *n*-alcohols in slab drying of aqueous maltodextrin solutions in dependence of drying temperature. Initial dissolved solids concentration 10 wt % (Rulkens, 1973).

His work showed that $D_{a,eff}$ increases with increasing initial water concentration and temperature, but is independent of other processing conditions. The effective diffusion coefficient again decreases with increasing aroma molecular mass. The length of the constant-rate period, which is the time in which the interfacial mass concentration of water decreases from the initial value $\rho_{w,0}$ to the critical value $\rho_{w,c}$ is determined by the steepness of the water concentration profile and by the concentration difference $(\rho_{w,o} - \rho_{w,c})$. The concentration gradient at the interface is given by:

$$-\partial \rho_w / \partial r = J_{w,i}/D_{ww} = \{ k' (\rho'_{w,i} - \rho'_{w,b}) \} / D_{ww} \quad (17)$$

The length of the constant-rate period t_c decreases with increasing steepness of the concentration profile, and thus with increasing k′ and decreasing bulk water concentration in the gas phase. Decrease in the initial water concentration will lead to a lower initial value of D_{ww} and a smaller difference $(\rho_{w,o} - \rho_{w,c})$ which both lead to a shorter time t_c. A decrease in initial water concentration also decreases $D_{a,eff}$, and thus will have a positive effect on AR. Increase in the initial thickness with other parameters constant increases t_c, but decreases t_c/R_0^2 (Kerkhof, 1974).

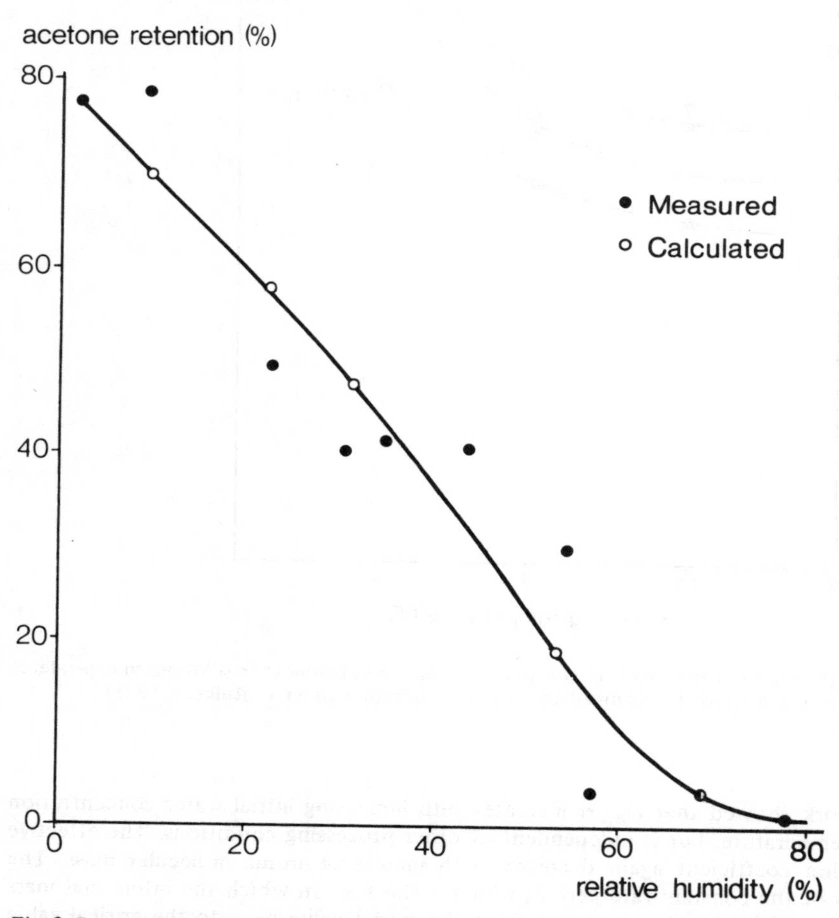

Fig. 9. Retention of acetone in slab drying of aqueous maltodextrin solution as a function of relative humidity of drying air (Menting, 1969).

The effect of temperature is more complicated. Increasing slab temperature leads to increasing $\rho'_{w,i}$, which tends to shorten t_c, but also increases D_{ww} and $D_{a,eff}$, which could increase t_c and Fo_c. Experimental data showed that aroma retention increased with temperature.

Figure 7 shows the aroma loss from drying gelled slabs of aqueous maltodextrin solution, as measured experimentally by Rulkens (1973) in relation to time, with the initial weight fraction of dissolved solids as parameter. An increase in aroma retention is clearly visible with increasing dissolved solids concentration and with increasing aroma molecular mass. Figure 8 presents the effect of slab temperature on aroma retention as given by Rulkens (1973). Increasing retention is observed with increasing temperature. Figure 9 presents the retention of acetone in gelled slabs of aqueous maltodextrin solution in relation to the relative humidity of the drying air, as found experimentally by Menting (1969, 1970a). It is clearly to be seen that aroma retention decreases with increasing relative humidity. Thus theory and experiment show that aroma retention increases with:

- increasing initial dissolved solids concentration
- increasing initial thickness
- decreasing relative humidity of the gas phase
- increasing mass transfer coefficient in the gas phase
- increasing temperature of the slab.

In conclusion, the diffusion model is an accurate description of the slab drying process, and the effect of process variables on aroma retention can be well understood from the theory based on this model.

Drying of solid droplets

The flux equations for water and aroma in the drying of solid spherical particles (see Fig. 2(b)), which are encountered in the first stage of spray drying and in extractive drying (Kerkhof & Thijssen, 1974) are identical to the ones for drying slabs. The diffusion equations for spherical coordinates read:

$$\partial \rho_w / \partial t = (1/r^2) \frac{\partial}{\partial r} (- r^2 J_w) \tag{18}$$

and

$$\partial \rho_a / \partial t = (1/r^2) \frac{\partial}{\partial r} (- r^2 J_a) \tag{19}$$

with the same initial and boundary conditions as for the drying slab.

The drying of slabs was treated for the isothermal case, which is justified as the experimental results known have been obtained for isothermal drying. The most extensive field of droplet drying is spray drying, which is definitely not isothermal for the droplet. Thus in the set of equations the heat balance should also be incorporated.

An other aspect typical for droplets in the coupling between mass transfer coefficient and droplet diameter, as given by the Ranz & Marshall relations (1952).

$$\mathrm{Sh} = k' 2R/D'_w = 2 + 0.6\, Re^{1/2} Sc^{1/3} \tag{20}$$

in which Sh = Sherwood number
Re = Reynolds number
Sc = Schmidt number
D'_w = diffusion coefficient of water in gas phase.

For small droplets moving not too rapidly, which is the case for droplets in spray driers at some distance from the atomizer the contribution of the Reynolds term can be neglected, resulting in

$$k' R = D'_w = \text{constant} \tag{21}$$

The non-isothermal drying of droplets was first calculated by van der Lijn (1972, 1975) for the drying of maltose solutions. Calculations including aroma transport were made by Kerkhof & Schoeber (1973, 1974), who also present a detailed review of theoretical aspects of spray drying. Concentration profiles calculated are similar to those in slab drying. Figure 10 shows the droplet temperature as function of time for two initial water contents for ideally mixed air flow. For both

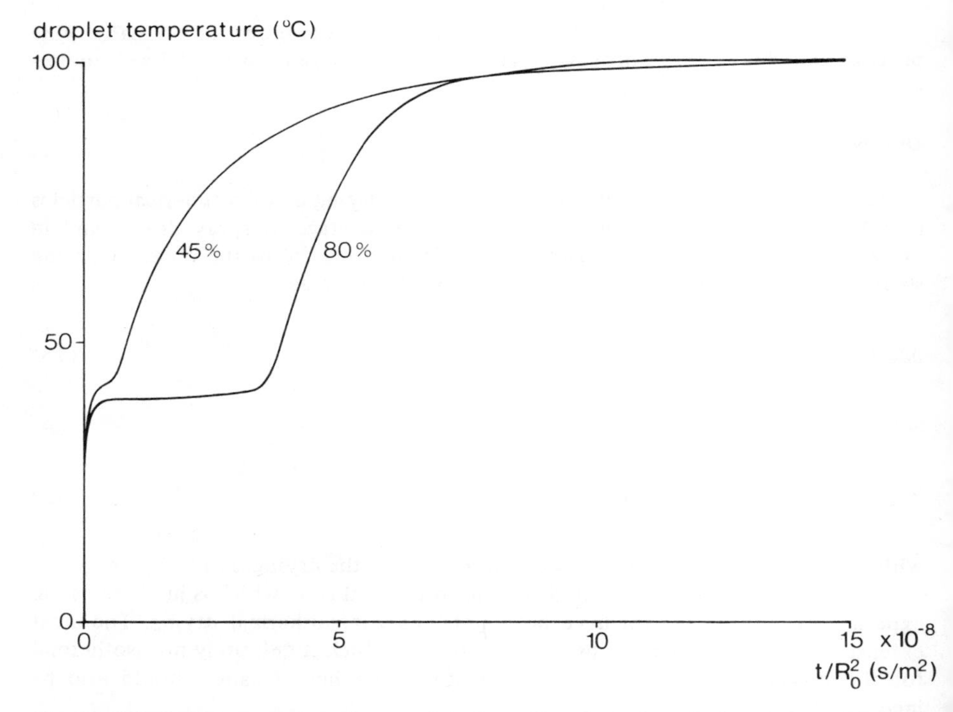

Fig. 10. Calculated temperature history of drying solid spherical particals at initial water contents of 45 and 80 wt % for aqueous maltose solution (Kerkhof & Schoeber, 1973; 1974). Air temperature 100 °C; air humidity 0.03 kg H_2O/kg dry air; air ideally mixed.

contents an interval of constant temperature is observed, during which the temperature remains at wet-bulb temperature. During this period, the water activity at the interface remains close to 1, and thus there is an equilibrium between evaporation of water and heat transfer. Upon passing the critical water concentration at the interface, the activity and consequently the evaporation rate of water decrease, and particle temperature increases. Analogous to the drying of slabs, the length of the constant-rate period decreases with decreasing initial water concentration. Similar curves were found experimentally by Menting (1969) and by Trommelen & Crosby (1970). Some calculated effects of process variables on aroma retention are present-

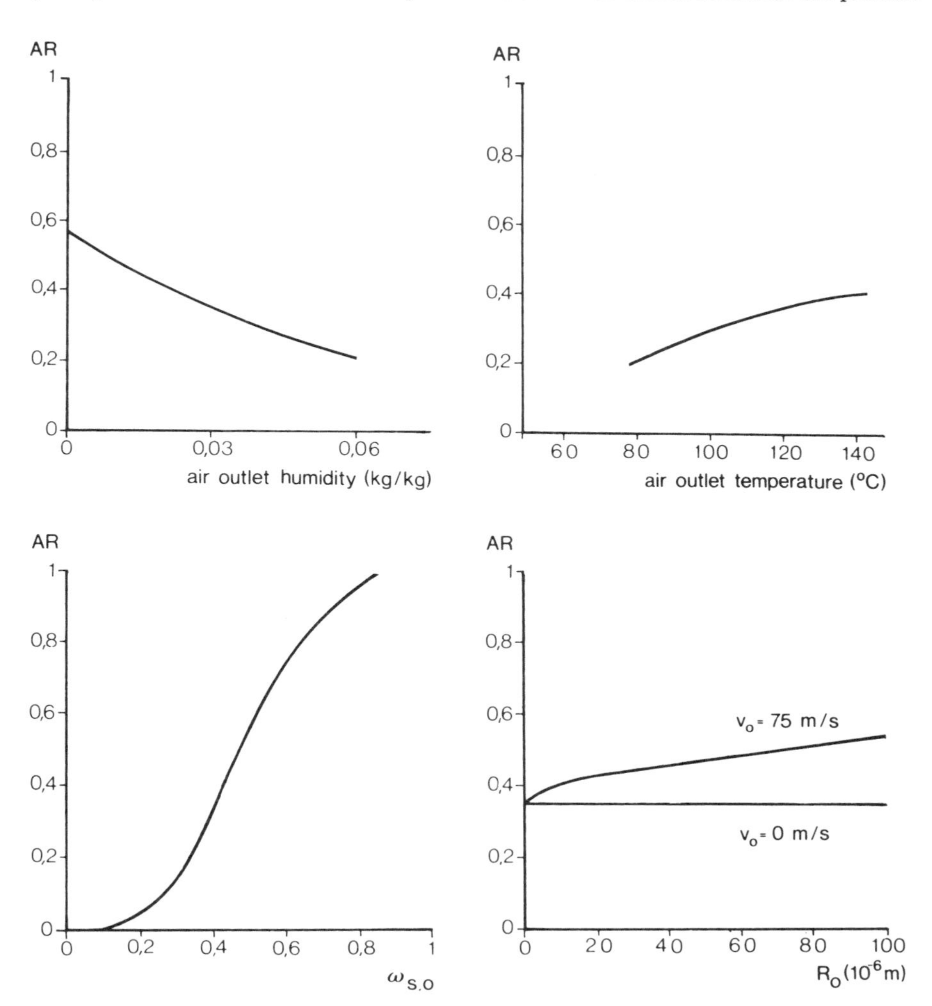

Fig. 11. Calculated effect of process variables on aroma retention in droplet drying (Kerkhof & Schoeber, 1973; 1974). Bottom left: initial mass fraction of dissolved solids. Bottom right: initial droplet radius.

ed in Figure 11. Increasing humidity increases the length of the constant-rate period, and thus decreases aroma retention. Increase in air temperature increases wet-bulb temperature, and thus aroma retention. Increase in the initial dissolved solids concentration decreases both the length of the constant-rate period and the effective diffusion coefficient of aroma and thus improves aroma retention.

Dimensional analysis (Kerkhof, 1974, 1975) shows that for relative velocity zero, t_c/R_0^2 is independent of R_0, but for high velocity t_c/R_0^2 decreases with increasing R_0. Thus at relative velocity zero aroma retention is independent of particle diameter, but it increases with R_0 for high velocities. In his experimental study of single droplet drying, Menting (1969) observed the same influences of process variables as calculated theoretically. He also observed, however, some additional causes of aroma loss. Upon increasing air velocity he found that internal circulation occurs at high velocity, increasing t_c and decreasing aroma retention. He also observed that at air temperatures above 100 °C the droplets puffed out, de-

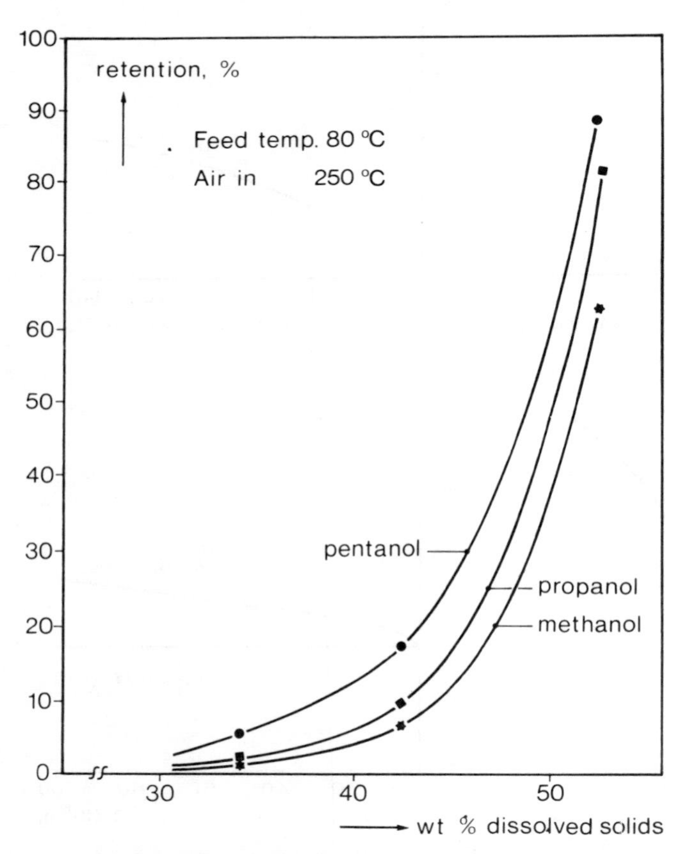

Fig. 12. Experimentally observed effect of feed concentration on aroma retention in spray dried maltodextrin solutions (Rulkens & Thijssen, 1972a). Temperature of feed: 80 °C; temperature of inlet air: 250 °C.

creasing aroma retention. Similar effects were observed in spray drying (Rulkens & Thijssen, 1972; Thijssen & Rulkens, 1968). They observed a strong increase in aroma retention with increasing initial dissolved solids concentration (Figure 12). Analogous results were found by Sivetz & Foote (1963), Brooks (1965) and by Reineccius & Coulter (1969). At increasing temperature they observed a decrease of aroma retention at high initial solids content. Their explanation, which was confirmed by Kerkhof (1975) later, was that with low dissolved solids content in the feed during the formation of the droplets and subsequent internal circulation, aroma loss is promoted by turbulent mass transfer inside the liquid phase, whereas at high initial dissolved solids content the particle expands and craters are formed upon overheating, thus rupturing the dry skin and leading to aroma loss from the particle interior. This will be treated in more detail in the next section.

Experiments on extractive drying were reported by Kerkhof & Thijssen (1974). In this process droplets of the liquid food were dispersed in a water absorbing liquid phase, such as polyethylene glycol (PEG). The main cause of aroma loss in this process was the deformation of the droplets upon entering the PEG. With a high viscosity of the feed it was found that very high aroma retentions could be achieved. Also in these experiments an increase of aroma retention was observed with increasing initial dissolved solids content.

Drying of hollow particles

As stated in the previous paragraph, the temperature of a drying particle in a spray-drier increases after the constant-rate period, by a lowering of the water activity at the evaporating interface. At low interfacial water activities, the temperature of the droplet approaches dry-bulb temperature. As the concentration in the centre may still be high, the equilibrium vapour pressure may well exceed atmospheric pressure, so that the droplet boils and craters form. This effect is stronger if small air bubbles are forced into the droplets during atomization. Since the inner atmosphere of an expanding particle consists of water vapour, there is no limitation for aroma transport by dry skin formation. By expansion, the diffusion distance to be covered by the aroma components becomes smaller too. Both factors may lead to considerable aroma loss. As, however, also the drying process proceeds much faster in this situation, the particle may be dry before the aroma is all lost. In atomization, particles vary in size, causing different drying times and trajectories through the spray-drier for particles of different size. As the spray drier is in general designed for the largest particles, the smaller particles will already be dry at the surface when only part of the path through the drier has been passed, and expansion occurs. The effect is stronger also at higher air temperature since then the constant-rate period is shorter and the boiling temperature is reached sooner. Since also for high dissolved-solids concentration the constant-rate period is very short, also in that case high expansion will be observed.

Therefore the factors promoting the expansion and consequent additional loss, are the same as those seen to improve retention of aroma in the solid particle. To exclude the unfavourable effect of expansion, spray-drying may be performed in two or more stages. This process, known as multiple-stage drying, or dual-drying, has recently been presented in various forms in the literature (Kjærgaard, 1974;

Bljumberg et al., 1970; Meade, 1971; Utag, 1971; Okada & Kato, 1972; Meade, 1973).

In this process the first stage consists of atomization and partial drying in hot air as in the conventional process. Then follow one or more stages such as bed driers or moving belt driers, in which the product is after-dried at lower temperature. Recently Kerkhof (1975) showed that indeed this technique may lead to higher aroma retention. By avoiding excessive particle temperatures, thermal deterioration can be suppressed too, as was reported for enzyme inactivation by Kjærgaard (1974).

Freeze drying of a slab

During freeze-drying the water is removed by sublimation from pure ice crystals, and evaporation of water from an interstitial lattice consisting of a highly concentrated amorphous solution. The water concentration in the concentrated solution at a given temperature is given by the freezing-point curve. The ice crystals are irregular and may differ considerably in size, as also are the interstitial lamellae. Thus exact modelling is impossible. However, a much simplified treatment may still provide insight into the effect of process variables on aroma retention. Figure 13 is a simplified diagram of freeze-drying of a slab. Let uniform ice crystals be assumed of diameter d_p, which after sublimation leave a pore of the same size. Further let the ice crystals be distributed regularly in the concentrated solution matrix, and let the average thickness of the interstitial walls be 2δ. Let further be assumed that after a given drying time, the ice front has retreated uniformly over the whole slab. This Uniformly Retreating Ice Front (URIF) model of Sandall et al. (1967) is well applicable in practice, as can be seen in extensive reviews by King (1971) and Karel (1973).

For the case of heat transfer through the dry layer,

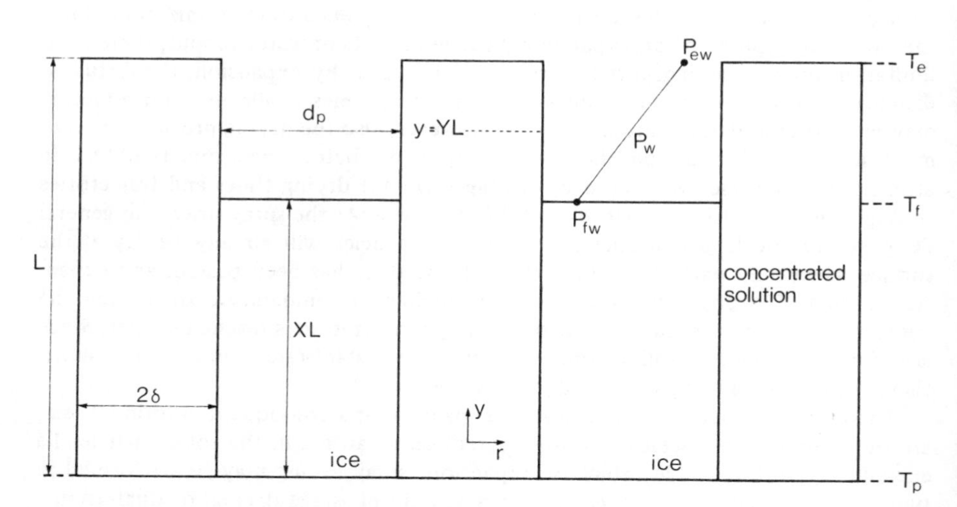

Fig. 13. Schematic representation of freeze drying of a slab.

$$q = \lambda\,(T_e - T_f)/L\,(1-x) \qquad (22)$$

in which
q = heat flux
λ = thermal conductivity of dry part
T_e = temperature at surface
T_f = temperature at ice front
L = height of slab
x = relative height of ice front.

For heat transfer through the frozen part,

$$q = \lambda_f\,(T_p - T_f)/Lx \qquad (23)$$

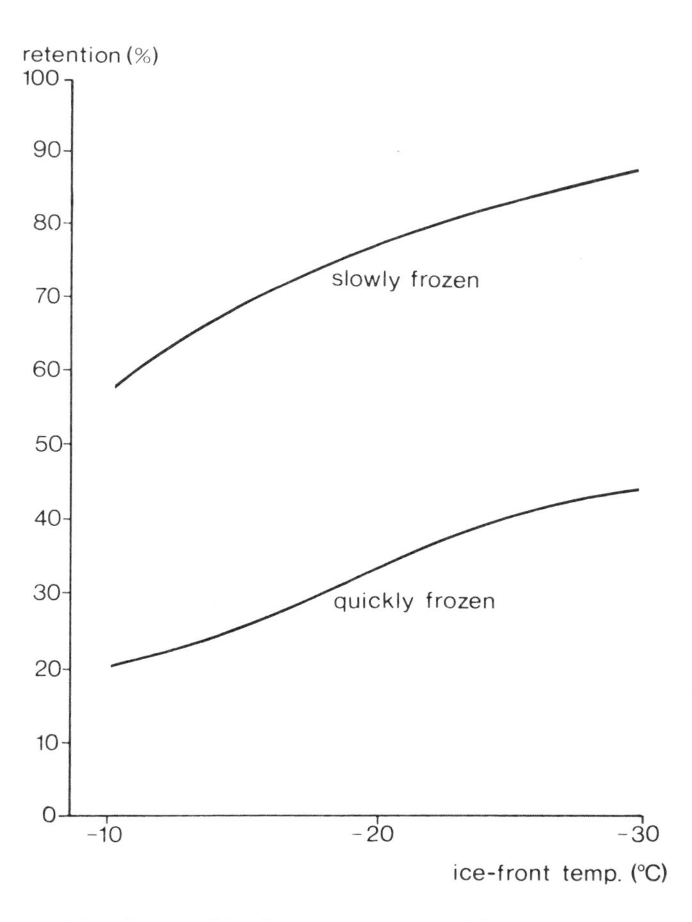

Fig. 14. Influence of ice front temperature and freezing rate on retention of acetone in freeze dried slabs of maltodextrin. Initial solids content 10 wt % (Rulkens & Thijssen, 1972b).

with λ_f = thermal conductivity of frozen part
T_p = platen temperature.

For the rate of ice sublimation,

$$N_w = \phi_{ice}\, \rho_{ice} \frac{dxL}{dt} = \left[\phi_{ice} K M_w \,/\, \{RTL(1-x)\}\right] (p_{fw} - p_{ew}) \tag{24}$$

in which
ϕ_{ice} = volume fraction of ice
K = permeability of dry layer
M_w = molecular mass of water
R = gas constant
T = absolute temperature
p_{fw} = vapour pressure at ice front
p_{ew} = vapour pressure in condensor room
ρ_{ice} = mass density of ice.

At low pressure, vapour flow can be described by Knudsen diffusion;

$$K = (2d_p/3)\,(2RT/\pi M_w)^{1/2} \tag{25}$$

Mass and heat fluxes are coupled by the relation

$$q = N_w \,.\, H_s \tag{26}$$

in which H_s = specific heat of sublimation, and by the vapour pressure relation:

$$p_{fw} = p_{fw}(T_f) \tag{27}$$

For not too large drying rates Rulkens (1973) analytically solved these equations, which for the total drying time τ can be written as:

transport through dry layer

$$\tau = \alpha\,(L^2/K) \tag{28}$$

transport through frozen layer

$$\tau = \alpha\,(L^2/K)\,(1 + \beta K) \tag{29}$$

in which $\alpha = (\rho_{ice} R T_f) \,/\, \left\{ 2M_w\,(p_{fw} - p_{ew}) \right\}$ (30)

and $\beta = (M_w^2\, H_s^2\, p_{fw}) \,/\, (R^2\, T_f^3 \lambda_{ice})$ (31)

Aroma retention is governed by the rate of aroma diffusion inside the interstitial liquid lamellae after the ice front has passed. As the wall thickness δ is much smaller than the slab height L, transport in the direction y may be neglected compared to that in the direction r. Numerical solution of the diffusion equations for water transport showed that there is virtually no concentration gradient of water in the r direction, and thus the water concentration is at equilibrium with local vapour pressure in the pore. As the aroma diffusion coefficient depends upon water concentration, the rate at which this diffusion coefficient decreases with time depends upon the rate of decrease in local water vapour pressure. This in turn is determined by the sublimation rate. Thus at any height in the slab, the time during which aroma is lost is coupled to total drying time. The rate of transport is further influenced by the wall thickness δ and by the initial diffusion coefficient of aroma $D_{a,o}$. Thus aroma retention can be coupled qualitatively to a Fourier number:

$$Fo = D_{a,o}\,\tau/\delta^2 \qquad (32)$$

and aroma retention will increase with decreasing Fourier number. For this Fo number can be written:

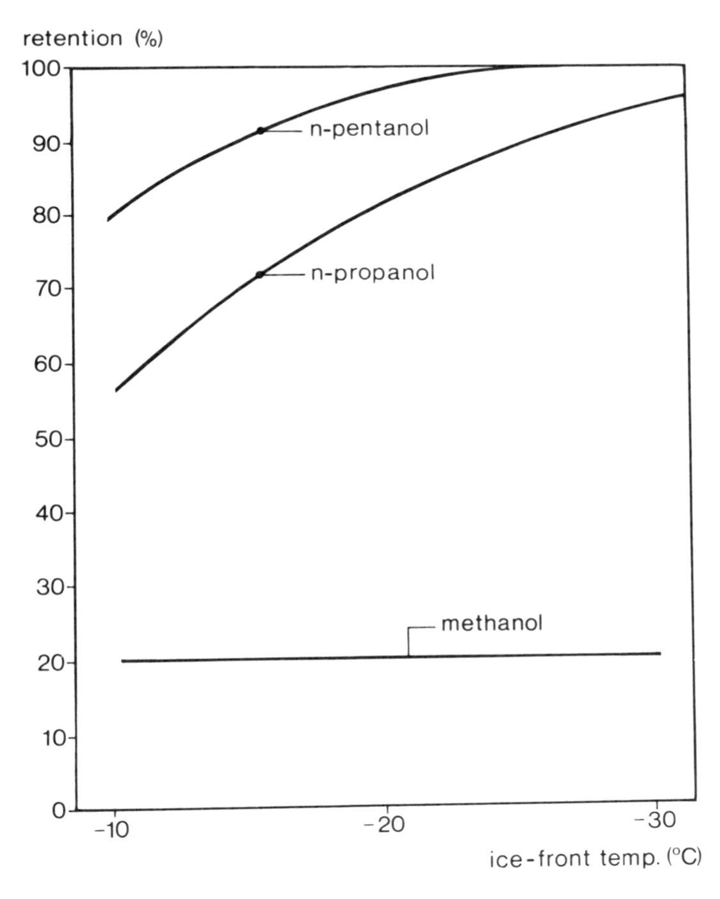

Fig. 15. Influence of molecular mass of aroma component and of ice-front temperature on aroma retention in freeze dried slab of maltodextrin (Rulkens & Thijssen, 1972b).

heat transfer through dry layer $Fo = \alpha (D_{a,o}/K)(L^2/\delta^2)$ (33)

heat transfer through frozen layer $Fo = \alpha (D_{a,o}/K)\{(L^2/\delta^2)(1+K)\}$ (34)

From this simple criterion the influence of process variables on aroma retention can easily be deduced:
– An increase in *freezing rate* decreases pore diameter d_p, and thus the permeability K and the wall thickness δ. Thus *Fo* will increase and aroma retention will decrease, as illustrated in Figure 14 (Rulkens & Thijssen, 1972b). This effect was also reported by many other authors (Lambert et al., 1973; Rulkens & Thijssen, 1969; Chirife et al., 1973; Flink & Karel, 1970b; Fritsch et al., 1971; Chirife & Karel, 1974; Capella et al., 1974; Voiley et al., 1971; Flink & Labuza, 1972; Petersen et al., 1973).
– An increase in *temperature at the ice front* decreases α, and increases K, β and $D_{a,o}$. The latter effect is very strong, since increasing the temperature also increases the water content of the matrix according to the freezing curve (Rulkens, 1973). As a result *Fo* increases and aroma retention decreases (Figure 14). Similar results were obtained by Saravacos & Moyer (1968) and Voiley et al. (1971).
– An increase in *molecular mass of the aroma component* in a homologous series is associated with a lower aroma diffusion coefficient $D_{a,o}$, and thus with lower *Fo* number and higher aroma retention (Figure 15; after Rulkens & Thijssen, 1972b). Similar results have been reported by Sauvageot et al. (1969); Capella et al. (1974) and Voiley et al. (1971). Again from Figure 15 the effect of ice-front temperature can be observed.
– A decrease in *chamber pressure* decreases α, and thus *Fo*, and consequently increases aroma retention. This has been observed experimentally by Rulkens & Thijssen (1972b), Capella et al. (1974), and Petersen et al. (1973).
– An increase in *layer thickness* increases *Fo* number and thus decreases aroma retention. Experimental evidence has been given by Rulkens (1973), Chirife et al. (1973), Flink (1969), Flink & Karel (1970b), and Flink & Labuza (1972).
– An increase in the initial dissolved solids concentration increases the thickness of interstitial walls, at approximately constant pore diameter. Thus with increasing initial dissolved solids concentration aroma retention increases, as was observed experimentally by Rulkens & Thijssen (1972b), Chandrasekaran & King (1970), Flink & Karel (1970b), Fritsch et al. (1971), Chirife & Karel (1974, 1975b), Ofcarcik & Burns (1974), Capella et al. (1974), Rey & Bastien (1962), Lambert et al. (1973).

Freeze drying of granules

In practice, liquid foods are granulated after freezing and thereupon freeze-dried; in most industrial applications, batch-tray freeze-driers are still used; recently, however, continuous freeze driers have been developed and applied (Lorentzen, 1974). Experiments of Rulkens and Thijssen (Thijssen, 1972; Rulkens, 1973) have shown that also in granule freeze-drying, aroma retention increases with increasing aroma molecular mass, with decreasing chamber pressure, increasing initial dissolved solids concentration and decreasing freezing rate, analogous to slab freeze

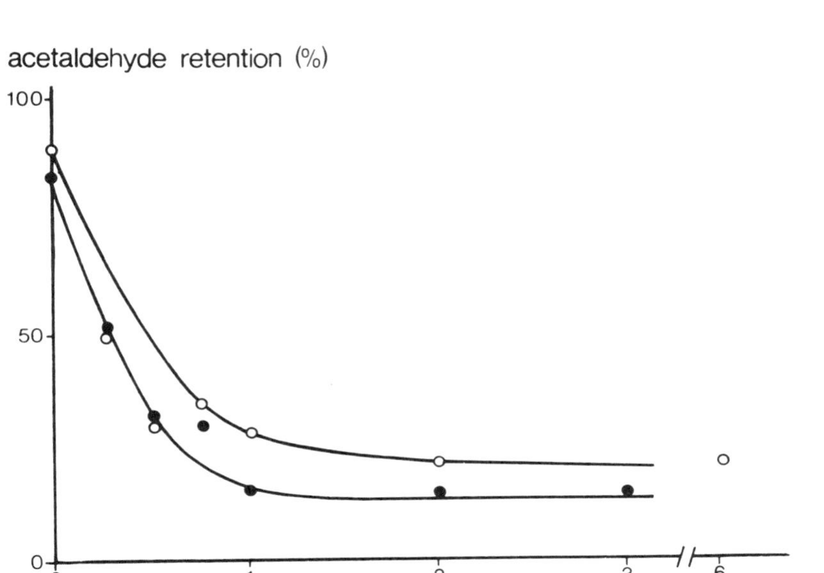

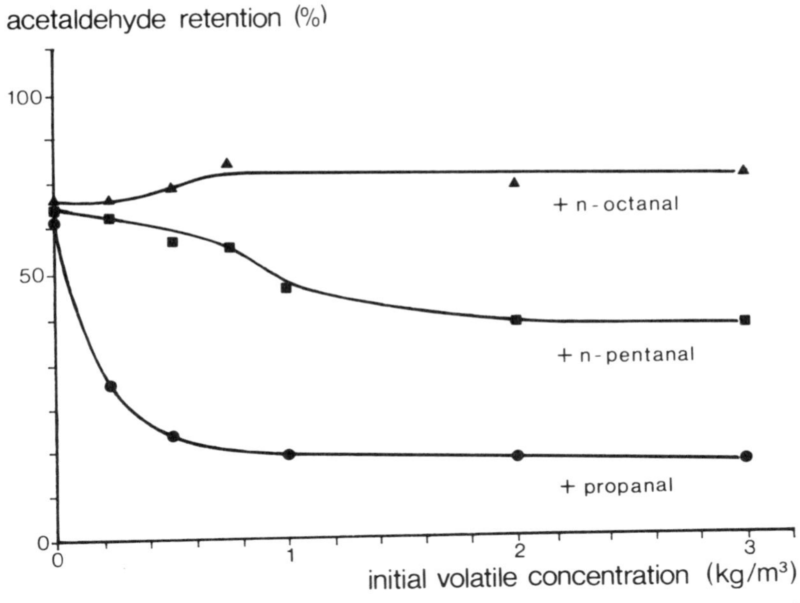

Fig. 16. Influence of addition of aldehydes on retention of acetaldehyde during freeze-drying of gum solution (Kayaert, 1974). Upper figure: ○ = initial acetaldehyde concentration 0.1 kg/m^3, ● = initial acetaldehyde concentration 1 kg/m^3.

drying. They also observed a decrease in aroma retention, AR, with decreasing particle size and with increasing layer thickness. Both effects increase resistance of the granule bed to water vapour flow and consequently decreases sublimation rate. Analogous to drying slabs, thus, the time during which aroma is lost increases and aroma retention decreases.

Effect of a dispersed phase upon aroma retention

In freeze-drying of model systems with model aroma components of limited solubility, an increase in initial aroma concentration decreases aroma retention (Flink, 1969; Massaldi & King, 1974a, 1974b; Kayaert et al., 1974; Kayaert, 1974). At higher concentrations, these aroma substances are present partly as dispersed droplets. Aroma retention decreases, with increase in aroma concentration because an increasing amount of aroma is present as droplets which evaporate upon contact with the open pores. Massaldi & King presented a model, which assumed that droplets in the model system adhere to the ice crystals in freezing and consequently directly evaporate after sublimation of the crystals. For orange juice, Massaldi & King observed that aroma droplets adhered to the cloud during freezing and drying and thus were retained inside the lamellae, in contrast to their results with model systems. Kayaert found, that considerable interaction occurred between different aroma components if at least one of the components has low solubility. Figure 16 shows the effect of addition of different aldehydes on retention of acetaldehyde. This effect still needs clarification.

The work of Kayaert and of Massaldi and King, and investigations in our own laboratory show that the essential factors in retention of an only slightly soluble aroma component are the size of the aroma droplets and of the lamellae, and the position of the droplets in the lamellae. These factors are influenced strongly by freezing conditions, as these determine droplet growth and droplet nucleation rate.

Investigations in our laboratory have shown that aroma retention for a dispersed aroma component increases with increasing initial dissolved solids concentration and with decreasing droplet size (Yamada et al., 1975).

Conclusions

The influence of process variables on the retention of aroma components in several drying processes leading to different product structures can qualitatively and, for simple geometries, be described quantitatively by selective diffusion. From the observed relations with process variables for each drying process, rules can be derived for obtaining optimum aroma retention.

Acknowledgments

The authors wish to express their gratitude to Mr H. S. Buytenhek for his valuable contribution to the literature survey.

References

Bartholomai, G. B., J. G. Brennan & R. Jowitt, 1974. 4th Int. Congr. Food Science and Technology, September 1974, Madrid.
Bartholomai, G. B., J. G. Brennan & R. Jowitt, 1975. Lebensm. Wiss. Technol. 8: 25.
Bljumberg, R. E., German Pat. Appl. nr. 1667205.
Bomben, J. L. & R. L. Merson, 1969. 66th Ann. Meet. AIChE, November 1969, Washington D.C.
Brooks, R., 1965. Birmingham Univ. Chem. Eng. 16: 11.
Capella, P., G. Lercker & C. R. Lerici, 1974. 4th Int. Congr. Food Science and Technology, September 1974, Madrid.
Chandrasekaran, S. K., 1969. Ph. D. Thesis, University of California, Berkeley.
Chandrasekaran, S. K. & C. J. King, 1971. Chem. Eng. Progr. Symp. Ser. 67 (108): 122.
Chandrasekaran, S. K. & C. J. King, 1972a. AIChE J. 18: 513.
Chandrasekaran, S. K. & C. J. King, 1972b. AIChE J. 18: 520.
Chirife, J. & M. J. Karel, 1973a. J. Food Sci. 38: 768.
Chirife, J. & M. J. Karel, 1973b. J. agric. Food Chem. 21: 926.
Chirife, J. & M. J. Karel, 1974a. J. Food Technol. 9: 13.
Chirife, J. & M. J. Karel, 1974b. Cryobiology 11: 107.
Chirife, J., M. J. Karel & J. M. Flink, 1973. J. Food Sci. 38: 671.
Flink, J. M., 1969. Ph. D. Thesis, Massachusetts Institute of Technology.
Flink, J. M. & M. J. Karel, 1969. 66th Ann. Meet. AIChE, November 1969, Washington D. C.
Flink, J. M. & M. J. Karel, 1970a. J. agric. Food Chem. 18: 295.
Flink, J. M. & M. J. Karel, 1970b. J. Food Sci. 35: 444.
Flink, J. M. & M. J. Karel, 1972. J. Food Technol. 7: 199.
Flink, J. M. & I. P. Labuza, 1972. J. Food Sci. 38: 119.
Fritsch, R., W. Mohr & R. Heiss, 1971. Chem. Ing. Tech. 43: 445.
Issenberg, P., G. Greenstein & M. Boskovic, 1968, J. Food Sci. 33: 622.
Karel, M. J., 1973. CRC Crit. Rev. Food Technol. 2: 217.
Karel, M. J. & J. M. Flink, 1973. J. agric. Food Chem. 21: 16.
Kayaert, G., 1974. Ph. D. Thesis, Cath. University Leuven. Published in Agricultura 22 (2).
Kayaert, G., P. Tobback, E. Maes, J. M. Flink & M. J. Karel, 1975. J. Food Sci. (in press).
Kerkhof, P. J. A. M., 1973. Informal Meeting 4th Int. Course Freeze Drying, June 1973, Bürgenstock.
Kerkhof, P. J. A. M., 1974. 4th Int. Congr. Food Science and Technology, September 1974, Madrid.
Kerkhof, P. J. A. M., 1975. Ph. D. Thesis, Eindhoven University of Technology, the Netherlands.
Kerkhof, P. J. A. M. & W. J. A. H. Schoeber, 1974. In: A. Spicer (ed.), Advances in preconcentration and dehydration of foods. Applied Science Publishers Ltd, London, p. 349.
Kerkhof, P. J. A. M. & H. A. C. Thijssen, 1974. J. Food Technol. 9: 415.
Kerkhof, P. J. A. M., W. H. Rulkens & J. van der Lijn, 1972. Int. Symp. Heat and Mass Transfer Problems in Food Engineering, October 1972, Wageningen, the Netherlands.
King, C. J., 1971. Freeze drying of Foods. CRC Press, Chemical Rubber Co., Cleveland.
King, C. J. & H. A. Massaldi, 1974. 4th Int. Congr. Food Science and Technology, September 1974, Madrid.
Kjaergaard, O. G., 1974. In: A. Spices (ed.), Advances in Preconcentration and Dehydration of Foods. Applied Science Publishers Ltd, London, p. 321.
Lambert, D., J. M. Flink & M. J. Karel, 1973. Cryobiology 10: 52.
Lijn, J. van der, W. H. Rulkens & P. J. A. M. Kerkhof, 1972. Int. Symp. Heat and Mass Transfer Problems in Food Engineering, October 1972, Wageningen, the Netherlands.
Lijn, J. van der, 1975. Ph. D. Thesis Agricultural University Wageningen, the Netherlands.
Lorentzen, J., 1974. In: A. Spicer (ed.), Advances in Preconcentration and Dehydration of Foods. Applied Science Publishers, London, p. 413.
Massaldi, H. A. & C. J. King, 1974a. J. Food Sci. 39: 438.
Massaldi, H. A. & C. J. King, 1974b. J. Food Sci. 39: 445.
Meade, R. E., 1971. Food Eng., July : 88.

Meade, R. E., 1973. Food Technol., December : 18.
Menting, L. C., 1969. Ph. D. Thesis, Eindhoven University of Technology, the Netherlands.
Menting, L. H. C., B. Hoogstad & H. A. C. Thijssen, 1970a. J. Food Technol. 5: 127.
Menting, L. H. C., B. Hoogstad & H. A. C. Thijssen, 1970b. J. Food Technol. 5: 111.
Ofcarcik, R. P. & E. E. Burns, 1974. J. Food Sci. 39: 350.
Okada, K. & F. Kato, 1972. U. S. Pat. nr. 3.596.699.
Petersen, E., J. Lorentzen & J. M. Flink, 1973. J. Food Sci. 38: 119.
Ranz, W. E. & W. R. Marshall Jr, 1952. Chem. Eng. Prog. 48: 141, 173.
Reineccius, G. A. & S. T. Coulter, 1969. J. Dairy Sci. 52: 1219.
Rey, L. R. & M. C. Bastien, 1962. Freeze Drying of Foods, National Academy of Science – National Research Council, Washington.
Rulkens, W. H., 1973. Ph. D. Thesis, Eindhoven University of Technology, the Netherlands.
Rulkens, W. H. & H. A. C. Thijssen, 1969a. Trans. Inst. Chem. Eng. 47: T292.
Rulkens, W. H. & H. A. C. Thijssen, 1969b. Rep. Int. Inst. Refrig. Comm. 10: 133.
Rulkens, W. H. & H. A. C. Thijssen, 1972a. J. Food Technol. 7: 95.
Rulkens, W. H. & H. A. C. Thijssen, 1972b. J. Food Technol. 7: 79.
Sandall, O. C., C. J. King & C. R. Wilke, 1967. AIChE J. 13: 428.
Saravacos, G. D. & J. C. Moyer, 1968. Chem. Eng. Prog. Symp. Ser. 64: 37.
Sauvageot, F., P. Beley, A. Marchand & D. Sinatos, 1969. Int. Inst. Refrig. Comm. 10: 133.
Schoeber, W. J. A. H., 1973. MSc. Thesis, Department of Chemical Engineering, Eindhoven University of Technology, the Netherlands. In Dutch.
Sivetz, M. & M. E. Foote, 1963. Coffee Processing Technology, AVI Publishing Company, Westport, Conn.
Thijssen, H. A. C., 1965. Inaugural Adress, Eindhoven University of Technology, the Netherlands.
Thijssen, H. A. C. & W. H. Rulkens, 1968. Ingenieur 80: CH 45.
Trommelen, A. M. & E. J. Crosby, 1970. AIChE J. 5: 857.
Utag G.m.b.H., 1971. Voedingsmiddelentechnologie 2 (50): 12.
Voiley, A., F. Sauvageot & D. Simatos, 1971. Progress in Refrigeration Science and Technology, 3, Proc. 8th Int. Congr. Refrigeration, Comm. 10, Washington.
Yamada, T., H. A. C. Thijssen & P. J. A. M. Kerkhof, 1975. To be published.

Proc. int. Symp. Aroma Research, Zeist, 1975. Pudoc, Wageningen.

Reaction of vanillin with albumin

Short communication

J. H. Dhont

Central Institute for Nutrition and Food Research TNO, Utrechtseweg 48, Zeist, the Netherlands

In this institute, experiments have been conducted on the relation between aroma components and the solid matrix of the food product. For the aromatization of 'synthetic' foods like those obtained from soya bean protein, we had the idea of adding stable and non-volatile derivatives of aroma substances to the product. By adding water to the product, the derivative was hydrolysed to the desired aroma substance and a non-toxic reaction product. We thought that in this way the product would combine a good shelf-life and the property of producing its aroma continuously for a certain time.

In one experiment with a stable derivative of benzaldehyde, we observed that the odour of benzaldehyde developed instantaneously on adding water to the derivative. If albumin was added to the derivative, it was some time before benzaldehyde could be smelled.

Chromatography of a mixture of vanillin and of albumin on a column of Sephadex, resulted in three instead of the expected two peaks. We substituted vanillin for benzaldehyde in this experiment because in some experiments benzaldehyde showed a peak attributable to benzoic acid formed by oxidation of benzaldehyde in air. Chromatograms from a typical experiment are shown in Fig. 1. The same phenomenon was observed with salicylaldehyde as test compound. There a yellow band was observed on the column between the protein and the aldehyde band.

In further experiments, we freeze-dried a solution of albumin and vanillin in a closed system. The condensed volatiles were collected in a cooled trap. After the experiment, the albumin residue had a deep yellow colour, which disappeared on solution in water. Extraction of the residue with ether removed the vanillin that was retained by the protein. The condensate was also extracted with ether and the vanillin content of both extracts was determined by ultraviolet absorption spectrometry. In this way, we found that the total amount of vanillin recovered by extraction was only about 10% of the amount added. With freeze-drying of vanillin solutions (without added albumin) 80–96% of the vanillin was recovered. So about 90% of the vanillin added must have been bound by the protein. In experiments with vanillin solutions, no vanillin could be found in the residue reservoir.

The distribution of vanillin originally added at the end of an experiment was: condensate 3, residue 8 and chemically bound 89%. Our experiments so far show that the protein retains some of the vanillin by encapsulation or adsorption, which

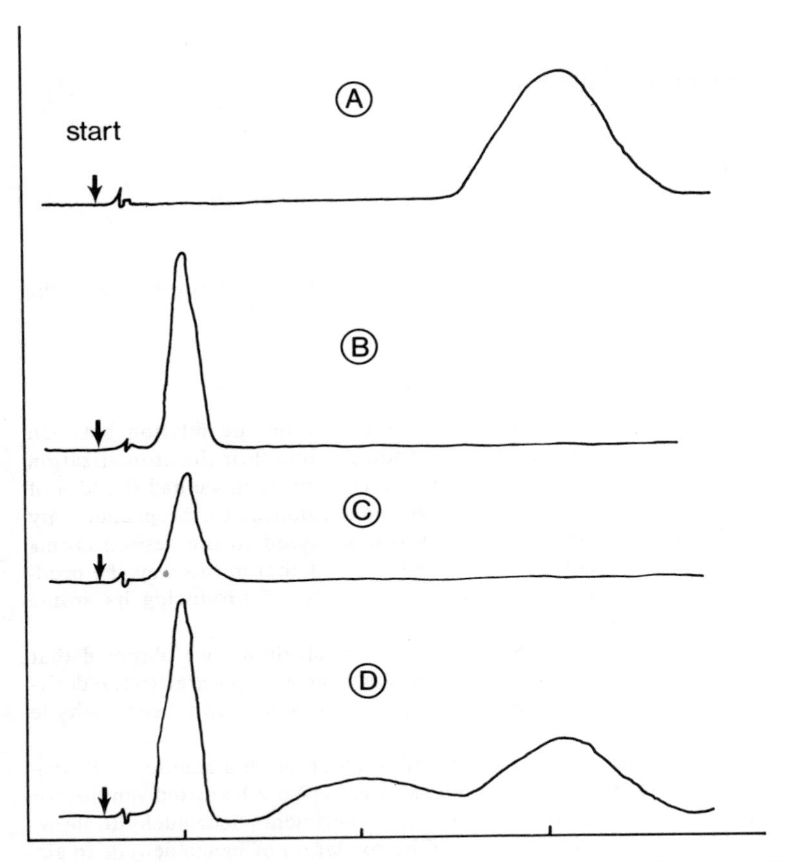

Fig. 1. Column chromatograms of A: 25 mg vanillin, B: 25 mg albumin, C: blue dextran, D: 25 mg albumin + 25 mg vanillin. Column: Sephadex G 25 (37 cm x 2.5 cm); eluent: water; flow: 3 ml/min.

contribute to the aroma over the product. However a much larger part of the vanillin is chemically bound to the albumin and does not contribute.

We suppose that the aldehyde group reacts with the primary groups in the protein. The reaction can be written schematically as follows:

$$RR_1C{=}O + H_2NR_2 \longrightarrow RR_1C{=}NR_2 + H_2O$$

Such reaction products are Schiff's bases. They are stable to alkali but are hydrolysed by strong mineral acids into the constituents.

29 May – The future of aroma research

Proc. int. Symp. Aroma Research, Zeist, 1975. Pudoc, Wageningen.

British and international legislative control of flavouring substances in food

T. J. Coomes[1]

Ministry of Agriculture, Fisheries and Food, Great Westminster House, Horseferry Road, London S.W. 1, England

Abstract

Most countries have no legislation on flavouring additives but for conventional additives usually have a positive list, which safeguards the consumer's health, taste and pocket. Despite differences from other additives, such a list. could also be made for approved flavourings. Approval could be based on need and safety-in-use. Such a list limits the liability of the manufacturer, protects the consumer and helps in international trade. Safety-in-use is difficult to define but criteria have been formulated for conventional additives by the FAO/WHO Joint Expert Committee on Food Additives and by national bodies. The United States Flavor and Extract Manufacturers Association and the Joint Expert Committee have also formulated criteria for flavourings, those of FAO/WHO being ultraconservative. A working party of the Council of Europe's Subcommittee on Health Control of Foodstuffs has produced a list of artificial flavourings (1) considered admissible, (2) temporarily admissible, and (3) not admissible at present. Their proposal is important in the evolution of consumer protection.

Flavouring agents fall clearly into the category of deliberate or intentional food additives, and in most countries of the world, such additives, with the exception of flavouring substances, are subject to legislative control. Such legislative control usually takes the form of an agreed positive permitted list, for food additives other than flavourings, after taking into account evidence of necessity, technical efficacy and safety-in-use. Perhaps it is pertinent to consider why such control of food additives is necessary in most countries of the world today. Great Britain's 1955 Food and Drugs Act (England and Wales) (GB, 1955) contains, in its general provisions, a general philosophy of control measures for foodstuffs sold for human consumption similar to those accepted by most Governments throughout the world, even though details of such legislation may differ somewhat, from country to country.

Section 1 of the 1955 Act makes it an offence to add any substance to food so as to render it injurious to the health of the consumer. Section 2 makes it an offence to sell food not of the nature, nor of the substance, nor of the quality demanded by the purchaser. Food bearing a name or description calculated to deceive the purchaser is prohibited from being offered for sale by Section 6 of the Act, whilst Section 8 renders it an offence to sell unsound food. Section 4 enables

1. The opinions expressed in this paper are the personal views of the author and not a statement of Ministry policy.

the responsible authorities to make specific regulations about food composition and its content of both deliberate and unintentional additives and contaminants. The total effect of similar legislation throughout the world may be summarized as protecting the health of the consumer, their palates, and their pockets!

Having considered some basic philosophy relevant to the legislative control of conventional deliberate food additives, let us ask ourselves if we wish to control flavouring substances in that way or at all? There are probably two or at most three main reasons for controlling flavouring substances. Firstly, a permitted list of substances which may be added to food, enshrined in law, in effect defines areas, within which the food manufacturer, or the seller of food, may operate without the burden of proof of safety-in-use resting on them. Secondly, such legislation defines areas within which the consumer may buy food, confident that the food is unlikely to be injurious to health when the additives are used under defined conditions. Thirdly, since there is a large measure of international agreement on the criteria desirable for establishing permitted lists of food additives, lists of substances could be harmonized between countries, allowing foods containing them to be sold across national boundaries, fostering increased trade and international prosperity.

If we accept that these are compelling reasons for legislative control of flavouring substances in food by procedures like those for control of deliberate food additives, the information required to enable the construction of a permitted list of flavourings may be summed up under two criteria, 'need', and 'safety-in-use'. Two queries spring to mind, if we think of flavourings as deliberate food additives. Firstly, do they differ in any way from additives already controlled by permitted lists, and secondly, should the criteria for flavourings be the same, or different from those for conventional food additives? There are no simple answers, but it is precisely the response to these fundamental queries that give some clues to meaningful control.

Conventional food additives, with the possible exception of colourings, fulfill clear-cut, if sometimes imperfectly understood, technical functions. In current sophisticated food technology, which has arisen to cope with the current population explosion, they are necessary. Flavourings exert less obvious effects, which must not be thought necessarily either deleterious or deceptive. The benefit to both palatability and physiological acceptability must be emphasized, bearing in mind that olfactory sensation has many interlinked components. Conventional food additives are few relative to flavourings. Whereas the number of food additives 'permitted' in some countries is of the order of some 250 substances, the number of individual flavouring substances used by industry is of the order of 3 000–4 000, so that flavourings outnumber their 'permitted' brethren by a factor of 10 or more. Some chemicals found to occur naturally in foods are used as additives in modern food processing. The great majority of flavourings occur in 'natural' products, many of which are used as foods. Lastly, the amount of flavouring substances in food is generally small relative to technical additives. Clearly, therefore, flavourings are different from conventional food additives. But does this mean that our criteria for assessment should also be different?

As in any modern legislation, we start from a status quo. Flavourings are used now, and have been used for many years in many countries throughout the world. Precedent is thus established, and existing usage has established need. Indeed, the

British Food Standards Committee said that, "if there is adequate evidence that the flavouring agents currently in use have no deleterious effects on health there could be no reasonable objection to their continued and widespread use" (GB-MAFF, 1965a). The criteria to be satisfied therefore relate to safety-in-use, and this has been a stumbling block for many attempts to list permitted flavourings in many countries throughout the world. Principal difficulties about assessment of safety in use are the large number of flavouring substances and materials, the scarcity of toxicologists and testing facilities throughout the world, and of course the cost of the work relative to the market or likely market for flavourings throughout the world.

A number of guidelines exist for establishing the toxicological properties of conventional food additives, as a result of work by the FAO/WHO (1958, 1962, 1966) Joint Expert Committee on Food Additives and its associated scientific Groups. In certain countries, national guidelines are available (e.g. GB-MAFF, 1965b). Although not all the available guidelines place equal emphasis on the various data specified as necessary for an evaluation, most of them agree on a comprehensive series of tests on laboratory animals including the following:

1. Acute tests, both oral and parenteral for rats, mice and usually a non-rodent.
2. Short-term feeding tests with several dietary levels on rats, mice and a non-rodent (90 days for rats and mice and 10% of the lifespan for other animals).
3. Long-term tests, including carcinogenicity to two species, extending over the lifetime of the animal (2 years for rats, 18 months for mice). Evidence on fertility is also desirable.
4. Evidence about the metabolism of the additive and where appropriate, its effect on enzymes.
5. Where such evidence can be obtained, effects on man are particularly appropriate and indeed WHO (1967) have drawn up principles for this type of investigation.

One must ask, therefore, whether such comprehensive requirements are justified or even necessary to enable assessments about safety-in-use of a flavouring substance. Toxicological hazards might arise from certain diets with large contents of flavouring agents. Animal experiments have shown us that some flavouring substances of natural origin are far from innocuous. Safrole from sassafras oil and coumarin from tonka beans are examples. Some substances, if ingested in minute amounts, can induce hypersensitivity as do capsaicin and zingiberin in some people or even allergic hypersensitivity as does menthol. Clearly, if any toxicologist be asked to decide whether traditional flavouring agents might be accorded some measure of favoured treatment, he does not have to look far to justify adoption of the usual criteria for assessment of conventional food additives.

However, since the late 1950s, there have been some examinations of the problem of assessing the safety-in-use of flavourings. The purpose of some was the construction of permitted lists of flavourings within the framework of both national and international food law. Thus, in 1959 the Flavor and Extract Manufacturers Association of the United States had to determine the status of flavouring substances that had been in use for a number of years at quite low concentrations in foods and beverages and set up a Scientific Panel to consider the problem. The FEMA panel worked out certain criteria which enabled a decision to be reached as to whether the use of a particular flavour could be regarded as toxicologically

insignificant. The conditions were:
a. use in food (including natural occurrence) for a period of at least 10 years without any evidence of adverse effect;
b. use in foods at contents not exceeding 10 mg/kg;
c. total annual use as a flavouring agent not exceeding 1000 lb (almost 500 kg);
d. simple chemical structure or composition whose metabolic fate is known or can be assumed to involve safe pathways.

If a flavouring substance complied with all four conditions, it could be regarded as having toxicologically insignificant usage.

FAO/WHO (1967) Joint Expert Committee on Food Additives examined a number of flavouring agents and always required or specified animal experiments for each substance under consideration, irrespective of any history of safe use. This situation, clearly ultra-conservative, was perhaps impractical and their recommendations for future action recognized the difficulty by stating that in future certain substances should have priority for toxicological evaluation. The Committee defined certain criteria for assessing priority:
a. appearance on existing restrictive positive lists;
b. estimated consumption per capita exceeding 3.65 mg/year;
c. use in foods at levels exceeding 10 mg/kg;
d. valid reasons for doubting safety.

Probably the most concerted and definitive approach to the problem of preparing permitted lists of flavourings for use in foodstuffs commenced when the Council of Europe's Sub-Committee on Health Control of Foodstuffs set up an ad hoc Working Party to study natural and artificial flavourings with the following aims (Council of Europe, 1965):
1. to draw up a list of the natural flavourings based on their sources and a list of artificial flavourings that may be added to foodstuffs without hazard to public health;
2. to draw attention to certain of those flavourings that present a hazard to public health;
3. to complete the list according to the procedure currently employed, using the information already collected and such information as could be supplied by interested parties;
4. to propose that from a certain date no new flavouring substance shall be used in foodstuffs unless the manufacturer supplies all the necessary data, in accordance with the common guide as set out, to prove it does not present a health hazard.

During the course of their studies, the Working Party took account of the following considerations:
a. certain natural flavourings are extracted from products which are normal items in human diet;
b. certain artificial flavourings are obtained by chemical processes; and
c. whilst the natural flavourings are mostly of a complex nature and cannot, in general, be precisely defined, most of the artificial flavourings have precise chemical specifications and can, therefore, be subject to purity standards and can be defined with more certainty.

In the present state of knowledge and for practical reasons, the Working Party considered it necessary to maintain a distinction between natural and artificial flavourings. They, therefore, drew up a list of flavouring substances which could be used in food, divided this list into two categories and sub-divided each of these categories on the following basis:

a. substances considered admissible;
b. substances considered temporarily admissible;
c. substances not considered admissible at present.

Lists (a) and (b) contain some flavourings for which specific limits are proposed. These limits were based provisionally on information available to the Council of Europe Working Party about amounts used in food, and, where available, the relevant toxicological data. The list of natural flavourings consists, essentially, of a list of botanical names of species, together with the parts of plants of those species, that may be used as flavourings or from which flavourings may be derived. The list of artificial flavourings consists of their chemical formulae and names (with synonyms).

The first version of the Council of Europe study which contained their recommendations for the control of natural and artificial flavouring substances in food by means of a permitted list was published in 1970 and circulated for urgent consideration by all interested parties. The comments received as a result of this publication have been further considered by the ad hoc Working Party and a second edition produced (Council of Europe, 1974). The preparation of a third edition is now in hand.

The work of the Council of Europe during the decade 1965–1975 on flavourings for use in food is an example of the evolution of public health proposals for consumers protection, utilizing existing information, both scientific and toxicological, from an enormous field within which the application of traditional criteria for such evaluations had so far even failed to establish guidelines.

As a further result of their definitive work, the Council of Europe ad hoc Working Party were able to construct a guide to the testing and toxicological evaluation of flavouring substances, including a series of conditions that must be satisfied before a flavouring substance can be accepted. I think it is true to say that, whilst the conditions are rigorous, they are not likely, except in special cases, to involve the formidable costs normally associated with complete evaluation of a traditional type food additive before its appearance in a list of permitted food additives.

Lastly of course, the assessment of safety-in-use aspects of flavourings are, as for conventional food additives, matters of opinion. Such technical, scientific and medical opinions are based on information available and the state of scientific knowledge advances daily, one is almost tempted to say hourly, and with it contemporary opinion. For that reason, current ideas on international legislative control of flavouring substances in food may also develop apace. To develop such ideas, however, will require a sound philosophy which has now been established.

References

Council of Europe, Subcommittee on Health Control of Foodstuffs, 1970–08–01. Natural and Artificial Flavouring Substances, SCHCF, Strasbourg.

Council of Europe, Subcommittee on Health Control of Foodstuffs, 1974. Natural Flavouring

Substances, their Sources, and Added Artificial Flavouring Substances, SCHCF, Maisonneuve.
Evaluation of the Carcinogenic Hazards of Food Additives, 1962. 5th Report FAO Nutrition Meetings Report Series 31.
GB-MAFF, Ministry of Agriculture, Fisheries and Food (England and Wales), 1965a. Food Standards Committee Report on Flavouring Agents. HMSO London.
GB-MAFF, Ministry of Agriculture, Fisheries and Food, 1965b. Memorandum on Procedure for Submissions on Food Additives and on Methods of Toxicity Testing. HMSO, London.
Great Britain: Parliament, 1955. Food and Drugs Act (England and Wales). 4 Eliz. 2. Ch. 16.
Procedures for the Testing of International Food Additives to Establish their Safety for Use, 1958. 2nd. Report FAO Nutrition Meeting Report Series 17.
Specifications for the Identity and Purity of Food Additives and their Toxicological Evaluation, 1968. 11th Report FAO Nutrition Meetings Report Series 44.
WHO, 1967. Technical Report Series 348.
WHO Scientific Group on Procedure for Investigating Intentional and Unintentional Food Additives in Order to Establish their Safety to the Consumer, 1966. Report to the Director General PA/66, 166, Geneva.

Proc. int. Symp. Aroma Research, Zeist, 1975. Pudoc, Wageningen.

The future of aroma research

F. Rijkens and H. Boelens

Naarden International NV, P. O. Box 2, Naarden/Bussum, the Netherlands

Abstract

An analysis of the future of aroma research starts with the evaluation of the history and the present state of our knowledge. The development of analytical aroma research is discussed and an assessment is given of the number of flavour components, yet unknown. Better guidance of analytical aroma research towards important flavour components is needed and ways to achieve this, for example by the use of sensorial evaluation techniques, are discussed.

Long term prospects about food production are expected to reflect further disconnection from the natural environment. Single cell proteins, plant tissue cultures and further growth of processed foods are foreseen.

Fundamental knowledge of biochemical and thermal flavour formation and of physico-chemical interactions of flavour compounds with non-volatile food components will be of growing importance.

Considering the invitation of the late Dr Weurman to discuss the future of aroma research, we realized that all of you, being deeply involved in flavour research, will have thought about this subject from time to time.

A paper on the future of aroma research cannot be expected to bring much new, but I hope I can provide some food for further thought. We will deliberately put some views and opinions provocatively for debate. Thus we will try to meet Dr Weurman's desire for a lively discussion.

History of aroma research

As a definition of aroma research, at least for the purpose of this paper, we have indicated the field in a slightly modified scheme given originally by Emily Wick (1965) (Fig. 1). I will not waste your time by putting it into words, just note that it includes perception of taste and odour and that we will not discuss texture although that too is an important element for the acceptability of food. Let us see where aroma research, thus defined, stands nowadays.

After one of the earliest successes of scientific research on aroma, the identification of vanillin from vanilla beans by Tiemann & Haarmann (1874), progress was at first slow. The isolation and identification of flavour components remained a very laborious and time-consuming operation, providing access mainly to major components. In the years up to 1950, there was considerable progress with the develop-

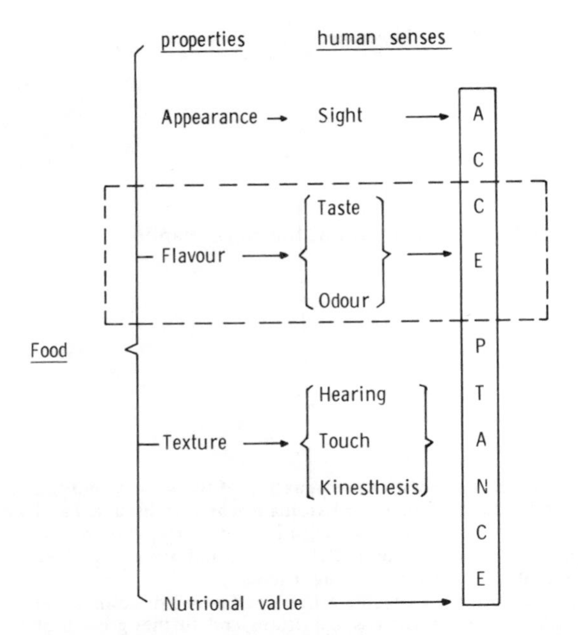

Fig. 1. Scheme of field of aroma research. After Wick (1965).

ment of organic chemistry, with better separation techniques like distillation and absorption chromatography, with better tools for identification such as micro-analysis of elements, and with ultraviolet and infrared spectroscopy.

But dramatic changes came with the development of gas chromatography and mass spectrometry. After the fundamental paper by James & Martin in 1952, gas chromatography came into widespread use since 1960–1965. This coincided with the development of not too expensive mass spectrometers, which brought this technique within reach of the analytical chemist.

These techniques and to a lesser extent proton magnetic resonance spectrometry have tremendously accelerated the output of analytical research on flavours.

I will not go into other stimulating factors such as the development of the food industry, the increased standard of living in many countries, the increasing demand for flavours or the growth of the flavour industry.

Recent progress and extrapolation to the future

The development and the present state of our knowledge of analytical aroma research can very well be demonstrated by the growth of the excellent 'Lists of volatile compounds in food' by Weurman and his successors, issued regularly since 1963 (van Straten & de Vrijer, 1973).

Fig. 2 shows the increase in number of *different* compounds over the years. The start of the accelerated growth rate around 1965 is clearly recognizable.

At present about 2600 different flavour compounds have been listed, or rather

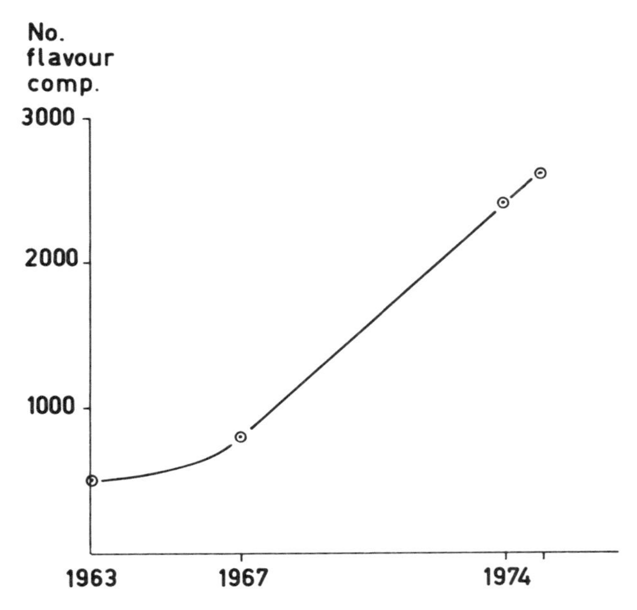

Fig. 2. Number of flavour components listed by Weurman (1963–1975). Generously provided by Dr Maarse, Central Institute for Nutrition and Food Research TNO, Zeist.

should I speak of more or less volatile compounds in foods since their real significance for flavours is not taken into account. A classification of these is given in Table 1. Strikingly low is the number of esters, 450, compared to the 230 acids known. What about the total number of 2600 flavour components and how many are there still to be detected in flavours? We will try to consider these questions.

Let us first discuss two different types of flavour, tomato and meat. Table 2 shows the flavour components of tomato identified in 5-years periods. In this paper I will use flavour components generally in a broad meaning, i.e. components that by their inherent properties *may* contribute to the flavour. After excellent work by several investigators like Viani et al. (1969), Schormüller & Kochmann (1969), Buttery et al. (1971) and others revealing the presence of over 250 components (Johnson et al., 1971), our colleagues Wobben et al. (1974) still found 63 compounds new for tomato flavour.

Figure 3 presents graphically the number of publications on tomato flavour and the number of components identified, against time. Obviously the number of papers annually on this subject is declining. As for the number of flavour components to be found, it does not seem unreasonable to expect that a total of between 400 and 500 components will be known in a few years.

Research on meaty flavours covers a much broader field. Table 3 summarizes the progress, again over 5-years periods. Here the output doubled in each 5-year period so that more than 500 components are now known, including a large proportion of nitrogen and sulphur compounds, about 20%.

Table 1. Classification of known flavour compounds.

	Number
Aliphatic/aromatic hydrocarbons	170
Isoprenoid hydrocarbons	130
Functionalized isoprenoids	170
Alcohols/phenols	190
Acetals/ethers	140
Carbonyls	310
Acids	230
Esters	450
Lactones	90
Furans/pyrans	110
Amino acids	40
Other nitrogen compounds	290
Thiazoles/oxazoles	60
Other sulphur compounds	220
Total	2600

Table 2. Tomato flavour components identified in 5-years periods.

Chemical compounds	1960 1965	1965 1970	1970 1975	Naarden 1974	Total
Hydrocarbons	0	27	2	5	34
Alcohols	8	15	15	4	42
Carbonyls	14	37	14	10	75
Acids	2	14	5	7	28
Hydroxy/keto acids	6	20	0	0	26
Esters	0	21	2	5	28
Lactones	0	5	1	6	12
Amino acids	0	0	4	0	4
Other nitrogen compounds	0	1	4	0	5
Sulphur compounds	0	3	1	1	5
Miscellaneous	4	28	18	25	75
Total	34	171	66	63	334

Extrapolation of this development to the future seems less certain than for tomato.

Figure 4 shows again a decline in the number of publications annually and therefore we can hardly expect doubling over the next 5 years (600 additional components, amounting to a total of 1100 components). However the number of components detected still rises sharply, leading us to the estimate that by 1980 another 300 components will be known, i.e. a total of about 800 components.

Both these examples illustrate the order of magnitude of the number of flavour components found when a flavour is analysed to the depth possible today. General-

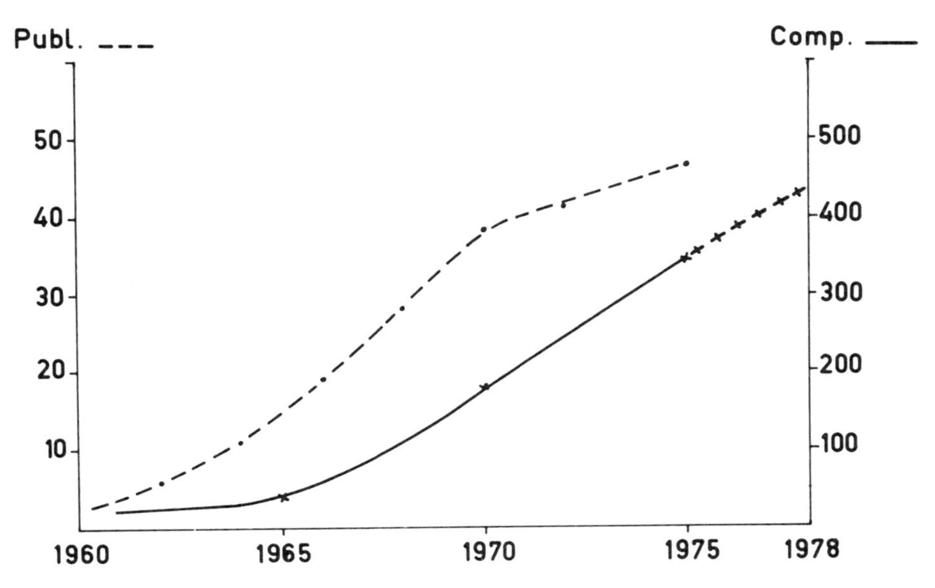

Fig. 3. Output of tomato flavour research. Left: number of publications. Right: number of compounds.

Table 3. Meat flavour components identified in 5-years periods.

Chemical compounds	1960 1965	1965 1970	1970 1975	Total
Hydrocarbons	0	34	32	66
Alcohols	2	19	14	35
Carbonyls	34	21	37	92
Acids	20	11	6	37
Hydroxy/keto acids	1	2	4	7
Esters	1	12	6	19
Lactones	0	1	14	15
Amino acids	0	25	8	33
Other nitrogen compounds	3	1	52	56
Sulphur compounds	6	15	24	45
Miscellaneous	4	17	104	125
Total	71	158	301	530

izing, a range from 300–800 seems a fair estimate since also less complicated flavours exist than meat or even tomato.

In our common food package, about 150 different types of flavour are present. Different varieties of one fruit like apple, or different types of whisky, cognac or wine are not regarded here as distinct types. Some 50 of these, about a third, at present have been investigated in appreciable detail. About 100 common flavour

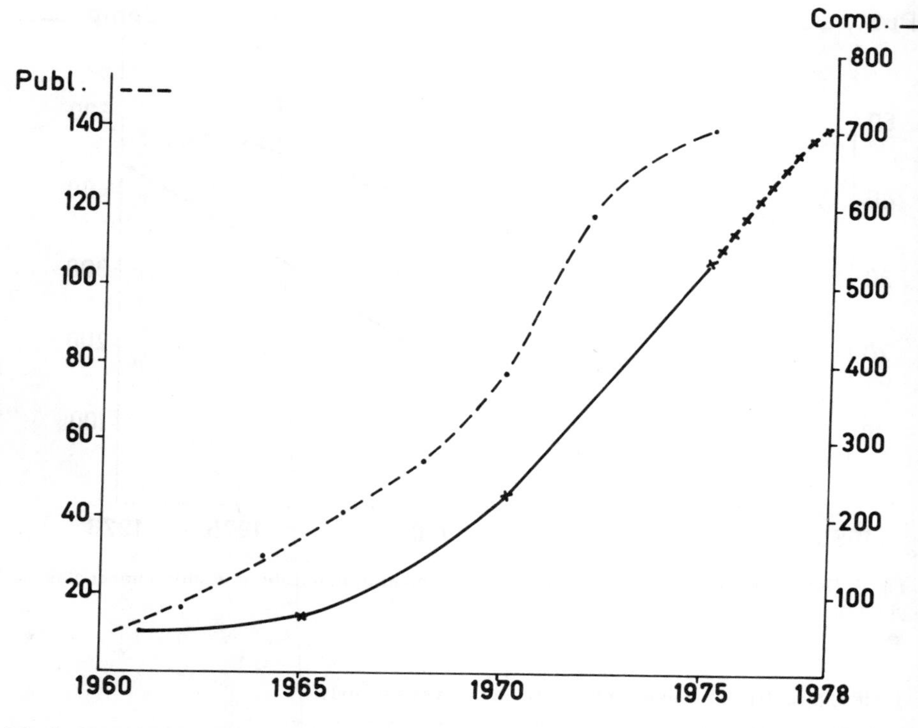

Fig. 4. Output of meat-type flavour research. Left: number of publications. Right: number of compounds.

Table 4. Number of expected flavour components; approximation.

Precursors	Number	Components[1]	Total
Amino acids	> 25	> 40	> 1000
Sugars	20	50	1000
Fatty acids	20	> 50	> 1000
Isoprenoids	150	5	750
Carotenoids	10	25	250
Benzenoids			500
Others:			500
– flavonoids/catechins			
– vitamins			
– alkaloids			
– nucleotids			
– steroids			
Total			> 5000

1. Per parent compound.

types and many others, less common, still remain to be investigated in detail and we are therefore convinced of the existence of at least double the number of flavour components presently known, let us say somewhere between 5 000 and 10 000.

This estimate can also be approached from another angle, from the flavour precursors (Table 4). From one amino acid, about 100 flavour components can be derived. From leucine, for example, 3 alcohols, 5 carbonyl compounds, 4 acids, 40 esters, 2 lactones, 25 acetals and some nitrogen and sulphur compounds, in total over 100 derivatives that in this case have actually been found. As a correction for overlap, we reduced the contribution of each amino acid to a safe minimum of at least 40. We know about 20 monosaccharides (C_4-C_7), stereoisomers excluded. Caramelization of glucose gives at least 100 flavour components. Including reactions with NH_3 and H_2S, the number of derivatives must be considerably larger. Therefore a count of 50 components per parent compound after correction for overlap is certainly not too high. From the fatty acids, only the natural linear unsaturated acids have been considered as precursors here. More than 20 of them are described. From one particular fatty acid about 10 carbonyl compounds can be derived by oxidation (oleic acid 7, linoleic acid 13, linolenic acid 19). These can form at least the same number of alcohols, of acids, esters and acetals each and additionally many other derivatives such as furans, lactones. Therefore 50 flavour derivatives per parent fatty acid certainly is a low estimate. We know about 50 monoterpenes and 100 sesquiterpenes. An assumption of the existence of 5 flavour derivatives (oxygen derivatives) per parent compound likewise seems not exaggerated. Finally we have added a rough estimate of the flavour derivatives from carotenoids, from cell wall material (benzenoids) and other precursors. Thus an analysis of the precursors also leads to the conclusion that a total of over 5000 flavour components must be expected.

Systematic analytical approach and alternatives

The following summarizes what we have said about the present situation in analytical aroma research:

– we know 150 common aroma types, of which 50 have been investigated in appreciable detail
– most aroma types contain between 300 and 800 components
– 2600 *different* flavour components are known; probably between 5000 and 10 000 exist.

A conclusion certainly could be that all this analytical work would take some 50 years and keep two generations of analytical flavour chemists busy. We are not going to wear ourselves out, I believe, by doing so systematically flavour by flavour. Our sharp analytical tools provide so many results, may of which are of limited value, that the need for guidance and limitation is clear. These have to come from a proper use of better sensory evaluation. Additionally, other approaches than systematic analysis should get more attention in aroma research. Further challenges will be provided by developments in food production and food technology. These will be important changes in future research on flavour, I believe.

Let us turn our attention now to these more motivating goals and opportunities. If we agree that many more different flavour components do exist, as yet unknown, then the intriguing question arises where and how these new compounds can be found. I will give a few examples.

The first day of this symposium we heard about the oxidation of linolenic acid in Grosch's excellent paper. Though a lot is known about the formation of saturated and unsaturated carbonyl compounds as the primary degradation products of linolenic acid, does the story end there? We do not think so. Depending on heat treatment, pH and other conditions, isomerization of the primary carbonyl compounds can lead to other α, β–unsaturated aldehydes (and 2,4-alkadienals) than those discussed. Moreover, the unsaturated carbonyl compounds may be oxidized further, without cleavage of the carbon chain, to the hydroxy and keto carbonyl compounds, which we do not know (Fig. 5).

2-Alkyl furans possibly may have been formed by allyl oxidation of 2-alkenals to 4-hydroxy-2-alkenals and then by cyclization. In the presence of hydrogen sulphide 2,4-alkadienals and 2-alkenals have been shown to form 2-alkyl thiophenes (Boelens et al., 1974) (Fig. 5).

Those 2-alkyl furans and the 2-alkyl thiophenes that have been found in cooked meat (Liebig et al., 1972; Wilson et al., 1973) and in cranberries (Anjou & von Sydow, 1967) could have been formed by these routes since the corresponding unsaturated aldehydes required as precursors have indeed been found in these foods (Rozier, 1970; Anjou & von Sydow, 1967). If this hypothesis be correct, the whole range of 2-alkyl furans and thiophenes corresponding to the unsaturated aldehydes present, may be expected in these and other foods.

We can also imagine (Fig. 5) the formation of

2-alkyl-2,3-dihydrofuran-3-ones
2-alkyl-2,3-dihydro-4-hydroxyfuran-3-ones

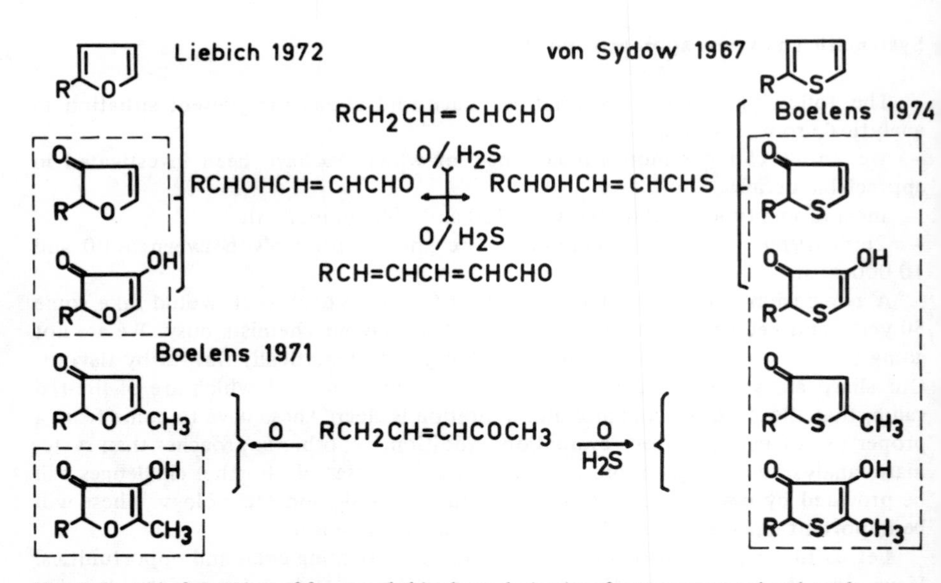

Fig. 5. Possible formation of furan and thiophene derivatives from unsaturated carbonyls.

2-alkyl-2,3-dihydrothiophen-3-ones
2-alkyl-2,3-dihydro-4-hydroxythiophen-3-ones.

2-*n*-Hexyl-5-methyl-2,3-dihydrofuran-3-one possibly derived from an unsaturated ketone has been found in onion oil (Boelens et al., 1971).

We are aware that some compounds of this type can also be derived from sugars (and hydrogen sulphide) (Schutte, 1974), but if unsaturated fatty acids could indeed act as precursors through the pathways suggested here, many new flavour components may be expected.

Another example of flavour formation is given in Fig. 6.

Our colleague (ter Heide, to be published) found in fenugreek hyperessence 3-methyl-γ-valerolactone, 2,4-dihydroxy-3-methylpent-2-enoic acid lactone (β, γ-dimethyl-α-tetronic acid) and the corresponding unsaturated amine. 2,4-Dihydroxy-3-methylpent-2-enoic acid lactone is a character-impact compound of fenugreek. Its possible origin from 4-hydroxyisoleucine lactone, an unusual amino acid found in fenugreek by Ghosal et al. (1974), is strongly suggested by the presence of the unsaturated amino lactone.

We did not search for the (normal) Strecker degradation products of the amino acids, but these compounds may be expected.

In addition to about 25 essential amino acids known, about 15 unusual amino acids have been found, but only a few of their normal degradation products (Strecker, Maillard) have yet been found. Therefore we feel the search for these compounds as new flavour components should be worthwhile. Particularly so since the more or less specific occurrence of unusual amino acids promises a higher chance that character-impact compounds be found among their derivatives.

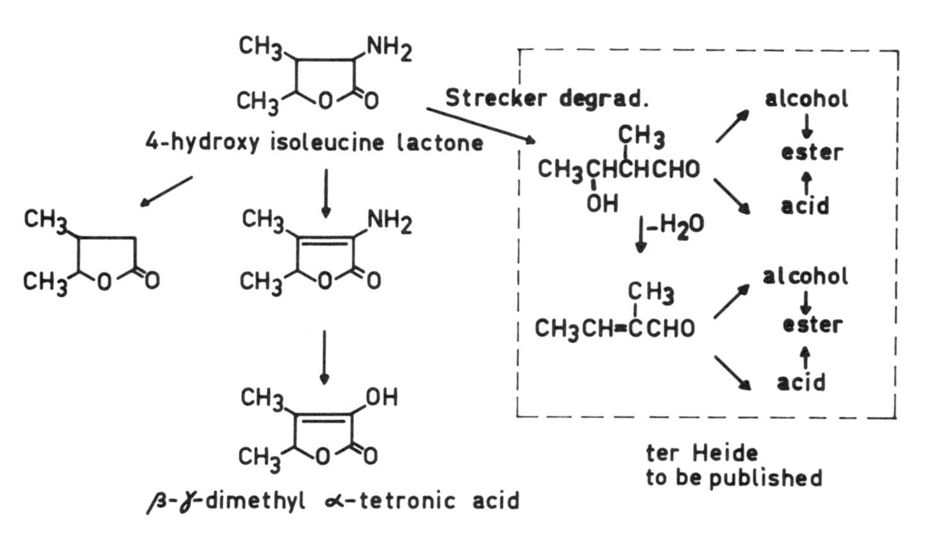

Fig. 6. Flavour components from an unusual amino acid isolated from fenugreek (*Trigonella foenum-graecum*).

Future aroma research and circumstantial trends

In the preceding part of this paper, we have considered what we called the systematic analytical approach to aroma research. But future progress is also dependent on more fundamental aroma research.

Fundamental chemical fields of aroma research are:

– the biogenesis of flavours
– the study of the thermal formation of flavours
– the study of physico-chemical interactions between volatile flavour compounds and non-volatile food components.

The aims of sensory research, of course, are in the first place the elucidation of:

– the mechanism of olfaction
– the mechanism of taste.

Further, attention should be paid to:

– sensoric evaluation methods
– specific areas like flavour enhancers and miracle fruit.

In this paper I cannot discuss, even briefly, the plurality of circumstances and developments that will influence the future of both fundamental and systematic flavour research, such as the economic situation, legislation, and health concern, including its rational and irrational aspects.

Let us limit our discussion to some main developments to be expected in food production and technology (Fig. 7). These will certainly stimulate and need funda-

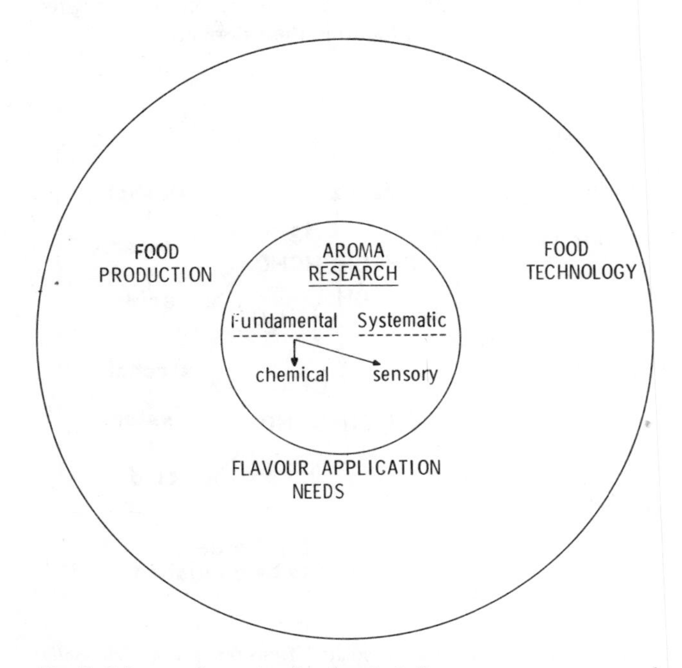

Fig. 7. Scheme of aspects of and main influences on aroma research.

mental flavour research and will continue to create problems requiring application of new knowledge of flavours.

The general trend will be that the production of foods will gradually become further disconnected from the natural environment and climate (e.g. Yoshikawa et al., 1974).

In fact this trend is clearly recognizable in many of past developments, although we do not realize so daily. The most recent examples of this trend are the bio-industrial production of poultry and pigs, and fish and seafood farming.

In the next decades, the *industrial* application of biological systems for food production will develop further, stimulated by the growing demand for food and made possible by the progress of the life sciences.

This sounds somewhat like science fiction, but I think it is not.

The most important biological systems for our food production are:

– plants and animals, mainly grown and reared in their natural environment, although industrial methods are gradually beginning to be applied.

– micro-organisms and fungi for the production of fermented foods and beverages, of protein hydrolyzates and proteins and of special metabolites like citric acid, glutamic acid, and lysine.

We have learnt in the past to disconnect micro-organisms from their natural environment and to grow them industrially for such purposes (and for other products like antibiotics). Having achieved this, we wanted to do more; we wanted to apply certain specific capabilities of these cells separately and developed processes to use micro-organisms for the production of

– enzymes. In the food industry, we all know the uses of single enzymes or mixtures, for example for the production of glucose and fructose from starch, to replace malt enzymes for beer and for the hydrolysis of sucrose. Proteases and lipases are used for the production of other food ingredients. We are learning now to immobilize enzymes so that they can be used repeatedly instead of just once.

The production of huge masses of cell material, necessarily connected with the production of desired metabolites, is becoming more and more a problem in waste disposal. Therefore research is being initiated to get rid of the cells and to achieve more complicated processes with their enzyme systems. One of the main difficulties here, however, is that we can only use enzymes efficiently in degradation processes since we have not yet learnt to feed in energy in a 'biological manner'.

I believe that the most logical future sequence of progress will be for us to learn how to grow plant cells on industrial scale and later perhaps animal tissue. At first, applications will be developed in which the whole cell mass is used. Later we will try to stimulate them to produce desired metabolites that can be isolated or render the whole cell mass more useful (Teuscher, 1973). Although the use of enzyme systems isolated from plants need not wait for industrial culture of plant tissue, it will certainly then be strongly stimulated.

The prospects for culture of plant tissue are very large indeed for several reasons and are no longer remote. It has been possible to propagate plant cells in artificial media and to grow complete plants in natural environment from small undifferentiated cell conglomerates. In fact this technique has been applied for several years by flower growers (Morel, 1964), and even on a large scale for the propagation of orange trees and pine trees.

More principal reasons for its potential, however, are the following aspects:
– spontaneous mutation occurs in propagation of plant cells for some generations, as in micro-organisms
– the possibility of manipulating plant cells genetically and chemically.

Therefore by appropriate selection of cells, we can obtain plant cells with even more useful properties than known varieties.

The application of plant tissue culture and selection will be extended first in agriculture for the development of higher-yielding and virus-free plants (Nishi & Ohsawa, 1973). The next major step will be the controlled differentiation of plant tissue and the growing of differentiated tissue of the required type, both in artificial environment. On laboratory scale, examples of these steps have been achieved and the production of tobacco plant tissue (Kobari et al., 1972; Shiio & Ota, 1973) seems even to have reached pilot plant stage. Then industrial photosynthesis with plant tissue (and with algae perhaps earlier) becomes possible.

There will be an increasing pressure for the production of carbohydrates because the need for proteins is gradually being satisfied. The application of soya protein grows and we will soon see the industrial production of single-cell protein for food from carbohydrate-rich waste products, including sulphite liquor or from petrochemicals as substrates (Worgan, 1974; Sherwood, 1974).

Recently the priority given to proteins as the most scarce food ingredient has been disputed and the world's needs for carbohydrates were deemed to become more critical (Yul, 1975).

The food processing industry will continue to expand production of processed foods based on raw materials like maize, other grains, potatoes, vegetable fat, and milk. The significance of such existing foods as industrial raw materials will continue to grow. Furthermore, the food processing industry will be an indispensable link in the utilization of future food ingredients like single-cell protein and products of industrial cultures of plant tissue.

It is obvious that the branches of fundamental aroma research are all of significant importance for these developments.

As several scientists have remarked (Nursten, 1970; Nursten, 1975; Solms, 1973; Rothe, 1974), a wealth of data on flavour components has been generated and this number will continue to grow considerably, but our knowledge of their relations and their origin is fragmentary. We cannot see the whole picture.

We are beginning to understand the biochemical pathways leading to terpenes, aliphatic alcohols, aldehydes, acids and esters or to the main components of some specific flavours such as those of onion and mustard seed. Flavour formation in apple or banana has been studied in considerable detail (Tressl & Jennings, 1972; Jennings & Tressl, 1974), but a general concept of the biochemistry involved in fruit ripening, for example, does not exist. For such a picture, much more biochemical investigation is needed, to clarify the relations between cell components, the biochemical capability spectrum of the cells and the changes in the biochemical metabolism inducing ripening and causing flavour formation.

Even less do we know the principles of thermal formation of flavours (Fig. 8), although these likewise are important in our diet. Many flavour components of chipped fried potatoes, meat and other heated foods are known and some pre-

Temperature: $100^{o}C$ ⟶ over $200^{o}C$

cooking, baking, (deep fat) frying, roasting.

Reactions of food components: cell wall material, polysaccharides, sugars, proteins, amino acids, fats, fatty acids, lecithins, carotenoids, vitamins, catechins, flavonoids, etc.

Other conditions: pH, cocatalysts like haem, access of oxygen.

Fig. 8. Scheme of thermal flavour formation.

Table 5. Composition of raw white rice and pork adipose tissue (%).

	Rice	Pork adipose tissue
Starch	80	0
Fat	0.3 – 0.5	90 – 95
Protein	5 – 10	0.5 – 2
Water	12	3 – 7
Ash	0.4	0.2

cursors have been identified, but a general picture of flavour generation from raw food components at different temperatures and under different conditions is far from clear.

As illustration I would mention some results on the formation of two flavours we obtained in close co-operation with the late Dr Weurman and Drs Schaefer of the Central Institute for Nutrition and Food Research TNO. This work deals with the flavours of cooked white rice and fried pork adipose tissue. The general composition of both raw foods is given in Table 5.

In the headspace of cooked rice, we found acetaldehyde, isobutanal and isopentanal and homologous series of the saturated aldehydes from pentanal up to undecanal and of the α,β-unsaturated aldehydes from 2-hexenal up to 2-decenal. We also detected hydrogen sulphide and ammonia and a trace of dimethyl sulphide.

To determine which are the precursors of the flavour volatiles, the fat fraction (0.3-0.5%) was isolated from uncooked rice by ether extraction. Of the fat fraction, 90% by mass consists of palmitic acid, oleic acid and linoleic acid, occurring for about 50% as the triglycerides. Upon boiling the fat fraction for 20 min with water, a headspace was formed containing the same spectrum of aldehydes as the headspace of cooked rice. The odour of the headspace was unspecific fatty and flourlike and was far from a good rice flavour.

Addition of appropriate amounts of hydrogen sulphide and ammonia to this headspace, however, yielded an almost complete rice flavour (Table 6).

These experiments show that the flavour of cooked white rice is mainly derived from the fat and the protein fractions. The essential part of the contribution of the protein fraction is the formation of hydrogen sulphide and ammonia.

In a similar investigation on lard flavour, the headspace of pork fat tissue heated at 170 °C, proved to contain the series of the saturated aldehydes from acetaldehyde up to and including nonanal, the α,β-unsaturated aldehydes acrolein up to and including 2-nonenal and hydrogen sulphide and methyl mercaptan. In further work designed to evaluate the role of different fractions obtained from adipose tissue, we found that homogenated tissue, after extraction with water, upon heating yielded the same spectrum of aldehydes, but yielded no sulphur compounds. It developed an unspecific bland fatty odour. Furthermore, the low-molecular, water soluble fraction to our surprise proved to generate a definitely lardlike flavour upon heating at 170 °C in hardened coconut fat. Since this fat in itself is practically inert, we concluded that the specific part of lard flavour originates from some of the water soluble components of adipose tissue: sugars, amino acids or oligopeptides (Table 7).

Future aroma research should rather emphasize the deeper study of facts like those just mentioned and the relative sensory importance of the flavour components formed. Such work will be fruitful, because it will provide an intelligent route to develop applications and will eventually lead to a thorough understanding of thermal flavour formation.

As a final remark about thermal flavour formation, we know little about flavour formation when water boils. This is particularly remarkable since we eat many cooked foods (potato, vegetables, rice, meat) and also since the flavour of many raw foods changes dramatically upon cooking, for instance in potatoes, asparagus, mushrooms, onions, or meat. This neglected field should get more attention.

Table 6. Cooked white rice flavour.

Main precursors	Main flavour contribution
Oleic acid (glyceride) Linoleic acid (glyceride)	saturated aldehydes (C_2; C_6-C_9) α,β-unsaturated aldehydes (C_6-C_8)
Amino acids (proteins)	H_2S, $(CH_3)_2S$, NH_3

Table 7. Lard flavour.

Main precursors	Main flavour contribution
Oleic acid (glyceride) Linoleic acid (glyceride)	saturated aldehydes (C_2-C_9) α,β-unsaturated aldehydes (C_3-C_8)
Amino acids Sugars Other water soluble compounds	H_2S, CH_3SH and carbonyl compounds specific aspects

Another important area for fundamental flavour research is the investigation of physico-chemical interactions between volatile flavour components and non-volatile food constituents. The practical significance of such interactions is clearly evident from the problems of flavouring soya protein. Work of Gremli (1974), Solms (1974) and others has shown that soya protein can firmly bind considerable amounts of certain flavour components, whereas it absorbs much less, or nothing at all, of others. As a result, many flavours change considerably or even disappear when brought into contact with soya protein.

Other food components like polysaccharides, starch for example, exhibit similar properties. Certainly connected with these phenomena are the spatial structures of proteins and polysaccharides. Therefore changes in their spatial structure may change the absorption of flavour components and thereby influence the perception of the flavour. I wonder whether changes may occur in the mouth during the chewing of food through salivary action, so influencing flavour perception. We know that saliva is capable of splitting glucosidic linkages of amylose thereby causing some sweetness, but as far as I know nobody has raised the question to what extent enzymic reactions during chewing are related to flavour perception.

Role of sensory research

I have touched several times upon sensory research. Of course its principal aim, the elucidation of the mechanism of olfaction and taste, is of outstanding scientific importance. It is difficult, however, to estimate the practical significance of our future understanding of these mechanisms. I am inclined to believe that for other fields of aroma research and application the most important aspect will perhaps be the understanding of interactions, i.e. the non-additive perception of flavour components. Before this goal is reached, however, aroma research can benefit from sensory research more than as yet. Aroma scientists should make more use of improved sensory evaluation methods. Perhaps we can use also results in some specific areas like flavour enhancers and miracle fruit.

We want to emphasize a few aspects of sensory evaluation that in our opinion are of importance. Until recently, the purpose of using panels mostly has been

- investigation of differences between people in perception of odour and taste
- investigation of consumer acceptance of complete flavours
- assessment and the quality control of the aroma of foods, flavours or flavour fractions.

Recently studies have been published that try to establish interrelationships between the chemical composition of a flavour and its sensory evaluation by a test panel meant to represent consumers. Suffice it to name a few workers only: von Sydow (1974), Bednarczyk & Kramer (1971), Moskowitz (1974), Powers & Quinlan (1974), Vuataz et al. (1974), Gostecnik & Zlatkis (1975).

It has become clear that evaluation of individual components or fractions as such can often be misleading because of non-additive effects. The combination of a few components can certainly evoke flavour characteristics not found in the separate components. We are convinced, also from sensory work on fragrances, that this principle should be formulated even more broadly, as follows. The effect of a single flavour component in a flavour is often closely determined by environmental conditions. In other words, the olfactive contribution of the same single component may vary from flavour to flavour. Therefore, we agree with the opinion that a reliable judgment of the relative importance of the individual contributions of single components or of fractions can best be obtained by comparing the complete flavour fortified with different amounts of the component to be evaluated with the flavour as such.

A useful tool in the analysis of the results of sensory evaluation by test panels is multiple regression analysis as a statistical method that correlates panel results with concentrations or peak areas in the gas chromatograms of the evaluated mixtures.

The application of the evaluation methods discussed does not require the participation of flavour chemists. Excellent results can be obtained with panels consisting of non-experts trained in the observation of flavours and flavour aspects and able to express their impressions adequately. The extended use and further refinement of methods for what has been called 'subjective-objective flavour evaluation' will be of great value for sensory research and will stimulate aroma research as well.

Conclusion

In this paper, we have listed our knowledge and we have outlined some areas of aroma research where much interesting and useful work can be done. Briefly:

- 150 common aroma types, 50 investigated in detail
- aroma types contain 300-800 components
- 2600 *different* components known, 5000-10 000 existing
- food production further disconnected from natural environment
- single-cell protein within 10 years
- industrial plant tissue cultures in next decades
- processed foods will grow further
- general concepts about biochemical and thermal flavour formation needed
- importance of flavour/substrate interactions
- better sensory evaluation required for faster progress.

We have discussed long-term prospects about our food production. For these, biochemical knowledge of flavour formation will be of increasing importance. The food industry will continue to provide new products like present-day convenience foods and will continue to use new food ingredients like soya protein, single-cell protein and other products of the search for new and improved food resources. In the processing of these to palatable food for the growing world population, flavour research will be an essential element. Certainly a large and worthwhile task.

References

Anjou, K. & E. von Sydow, 1967. The aroma of cranberries. 2. Vaccinium macrocarpon. Acta chem. Scand. 21: 2076–2082.

Bednarczyk, A. A. & A. Kramer, 1971. Practical approach to flavor development. Food Technol. 25: 1098–1107.

Boelens, M., P. J. de Valois, H. J. Wobben & A. van der Gen, 1971. Volatile flavor compounds from onion. J. agric. Food Chem. 19: 984–991.

Boelens, M., L. M. van der Linde, P. J. de Valois, H. M. van Dort & H. J. Takken, 1974. Organic sulfur compounds from fatty aldehydes, hydrogen sulfide, thiols and ammonia as flavor constituents. J. agric. Food Chem. 22: 1071–1076.

Buttery, R. G., R. M. Seifert, D. G. Guadagni & L. C. Ling, 1971. Characterization of additional volatile components of tomato. J. agric. Food Chem. 19: 524–529.

Ghosal, S., R. S. Srivastava, D. C. Chatterjee & S. K. Dutta, 1974. Fenugreekine, a new steroidal sapogenin-peptide ester of Trigonella foenum-graecum. Phytochemistry 13: 2247–2251.

Gostecnik, G. F. & A. Zlatkis, 1975. Computer evaluation of gaschromatographic profiles for the correlation of quality differences in cold pressed orange oils. J. Chromatogr. 106: 73–81.

Gremli, H. A., 1974. Interaction of flavor compounds with soy protein. J. Am. Oil Chem. Soc., special issue 51: 95A–97A.

Grosch, W., 1975. Aroma compounds formed by enzymatic co-oxidation. Proc. int. Symp. Aroma Research, Zeist, 1975. Pudoc, Wageningen, p. 248.

Heide, R. ter. To be published.

James, A. T. & A. J. P. Martin, 1952. Gas-liquid partition chromatography; the separation and microestimation of volatile fatty acids from formic acid to dodecanoic acid. Biochem. J. 50: 679–690.

Jennings, W. G. & R. Tressl, 1974. Production of volatile compounds in the ripening Bartlett pear. Chem. Mikrobiol. Technol. Lebensm. 3: 52-55.

Johnson, A. E., H. E. Nursten & A. A. Williams, 1971. Vegetable volatiles: a survey of components identified. Part 2. Chem. Ind. 1212–1224.

Jul, M., 1975. Natural proteins as food components. TNO meeting. Effects on industry of trends in food production and consumption, Rotterdam 1975. Chem. Weekbl. 71 (10) :5.

Kobari, M., H. Kagiwara & K. Uchiyama, 1972. Culturing leaf tobacco for making smoking products. Fr. 2. 107. 371. Chem. Abstr. 78: 13847.

Liebich, H. M., D. R. Douglas, A. Zlatkis, F. Müggler-Chavan & A. Donzel, 1972. Volatile components in roast beef. J. agric. Food Chem. 20: 96–99.

Morel, G. M., 1964. Tissue culture; a new means of clonal propagation of orchids. Am. Orchid Soc. Bull. 33: 473-478.

Moskowitz, H. R., 1974. Combination rules for judgments of odor quality difference. J. agric. Food Chem. 22: 740–743.

Nishi, S. & K. Ohsawa, 1973. Mass production method of virus-free strawberry plants through meristem callus. Jap. agric. Res. Quart. 7: 189–194.

Nursten, H. E., 1970. Volatile compounds; the aroma of fruits. In: A. C. Hulme (Ed.), The biochemistry of fruits and their products, Vol. 1, p. 239–268.

Nursten, H. E., 1975. Chemistry of flavours; past, present and future. Int. Flavour Food Add. 6 (2): 75–82.

Powers, J. J. & M. C. Quinlan, 1974. Refining of methods for subjective-objective evaluation of flavor. J. agric. Food Chem. 22: 744–749.

Rothe, M., 1974. Aufgaben und Probleme der modernen Aromaforschung. Nahrung 18: 115–123.

Rozier, J., 1970. La flaveur des viandes et des produits de charcuterie. Parfums Fr. 13: 244–251.

Schormüller, J. & H. J. Kochmann, 1969. Gaschromatographisch-massenspectrometrische Analyse leichtflüchtiger Aromastoffe der Tomate. Z. Lebensm-Untersuch. Forsch. 141, Heft 1: 1–9.

Schutte, L., 1974. Precursors of sulfur-containing flavor compounds. Crit. Rev. Food Technol. 4: 457–505.

Sherwood, M., 1974. Single-cell protein comes of age. New Sci. 28 Nov.: 634–639.

Shiio, J. & S. Ota, 1973. Alkaloid production by tissue culture. Japan. Kokai 73.91287. Chem. Abstr. 80: 118396.

Solms, J., 1973. Die Bedeutung von Lebensmittelaromen. Chem. Rundsch. 26 (15) :19.

Solms, J., 1974. Flavors and the nonvolatile components of foods. Atti conv. Ital. sostanze aromatizzanti origine vegetale alim. mod. 1: 56–59.

Straten, S. van & F. de Vrijer, 1974. Supplement to the lists of volatile compounds in food; 3rd edn., 1973. Report R 4030, Central Institute for Nutrition and Food Research TNO, Zeist, the Netherlands.

Sydow, E. von, 1974. Can a sensory panel be replaced by an instrument. Inst. Food Sci. Technol. Proc. 7: 190–192.

Teuscher, E., 1973. Probleme der Produktion sekundärer Pflanzenstoffe mit Hilfe von Zellkulturen. Pharmazie 28: 6–18.

Tiemann, F. & W. Haarmann, 1874. Ueber das Coniferin und seine Umwandlung in das aromatische Princip der Vanille. Ber. Dtsch. chem. Ges. 7: 608–623.

Tressl, R. & W. G. Jennings, 1972. Production of volatile compounds in the ripening banana. J. agric. Food Chem. 20: 189–192.

Viani, R., J. Bricout, J. P. Marion, F. Müggler-Chavan, D. Reymond & R. H. Egli, 1969. Sur la composition de l'arôme de tomate. Helv. chim. Acta 52: 887–891.

Vuataz, L., J. Sotek & H. M. Rahim, 1974. Profile analysis and classification. Abstr. papers 4th int. Congr. Food science and technology, September 1974, Madrid, 1a. 10.

Wick, E. L., 1965. Chemical and sensory aspects of the identification of odor constituents in foods; a review. Food Technol. 19: 827–833.

Wilson, R. A., C. J. Mussinan, I. Katz & A. Sanderson, 1973. Isolation and identification of some sulfur chemicals present in pressure-cooked beef. J. agric. Food Chem. 21: 873–876.

Wobben, H. J., P. J. de Valois, R. ter Heide, H. Boelens & R. Timmer, 1974. Investigation into the composition of a tomato flavour. Abstr. papers 4th int. Congr. Food Science and Technology, September 1974, Madrid, 1a. 4.

Worgan, J. T., 1974. Single cell protein. Plant Foods Man 1: 99–112.

Yoshikawa, S., S. Kimura, S. Nishimaru, M. Miyuzaki, S. Tamura, K. Katoh & N. Ishima, 1974. Delphic studies on future eating life and research needs. Round table meeting 4th Int. Congr. Food Science and Technology, September 1974, Madrid.

Proc. int. Symp. Aroma Research, Zeist, 1975. Pudoc, Wageningen.

Heterocyclics in flavour chemistry

Five and six-membered rings containing oxygen, nitrogen and sulphur atoms; monocyclic and condensed bicyclic components

I. Flament

Firmenich SA, Research Division, P.O. Box 239, CH-1211 Geneva 8, Switzerland

Abstract

Heterocyclics of food aromas are classified according to the dimension of the ring and the number of heteroatoms. Each skeleton is illustrated by the most characteristic coumpounds, either with the basic structure or with typical organoleptic properties. The number in certain classes (such as furans) has necessitated subdivision into further classes, thereby allowing the variety of functional groups encountered in natural compounds to be demonstrated. In addition, some artificial heterocyclic structures are given. They are divided into deduced and related structures in decreasing order of likelihood of occurence in a natural flavour.

Among the known volatile components of food flavours, heterocycles are important. They are numerous, frequently abundant and present a high variety of interesting organoleptic properties. Our intention is to classify these chemicals according to the dimension of the ring and the number of heteroatoms, disregarding the substituents and the degree of oxidation of the ring itself. We intentionally left aside bridged and spiran structures, and non-volatiles (flavour potentiators) although important in flavour properties of certain foods.

Natural and artificial heterocyclics

Systematic exploitation of analytical results started 20 years ago with the advent of modern instruments or processing equipment. Since then, thousands of volatiles have been identified in a few hundred natural products. But even in the nineteenth century, pioneers painstakingly isolated and identified components whose organoleptic interest was immediately recognized. For example, Ciamician & Silber isolated the phthalides from celery oil, and elucidated their structure in 1897. Other heterocyclics had a much slower start: Schrötter (1879) isolated some alkylpyrazines from a fusel oil, and Reichstein & Staudinger (1926) found them in various alcoholic beverages and even in coffee aroma, without recognizing their interest. Only during the last ten years have their wide distribution and importance in roasted flavours been realized.

Since the 1960s, much analytical and synthetic work on flavour chemistry has been published. The low detection limits of modern chromatographic techniques have allowed the discovery of numerous components present only in traces.

But chemists had wider ambitions than mere analysis: they have used their reasoning powers and have synthesized products that 'should be present' in nature. The first method of reasoning is to classify compounds with a flavour to make a comprehensive survey. By analogy, one can then deduce analogous structures and synthesize compounds to fill gaps in the systematic classification. These 'deduced structures' may be present in natural flavours. Their identification is then only a question of time. An example will be detailed later.

A second method of anticipating structures of unidentified compounds is based

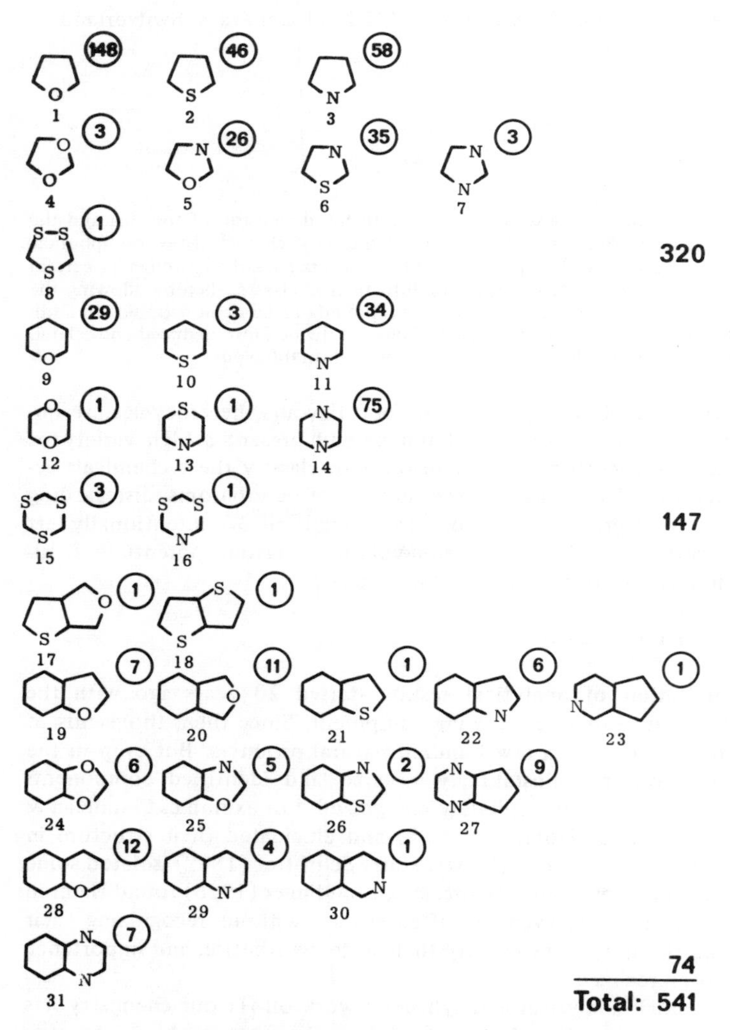

Fig. 1. Heterocyclic skeletons. The figures in circles represent the identified number of natural aroma substances with the skeleton. Totals for groups are given on the right.

on model reactions. A good example is the study of meat flavour with roasting. Amino acids present in water-soluble precursors can be pyrolysed with or without a reducing sugar and the products can be examined. Although they cannot be considered as natural compounds, they might some day be identified in a flavour whose complexity defeats analysis.

A more sophisticated approach, aiming to prepare compounds with interesting organoleptic properties, is to modify the structure of natural products either by extending the length of a chain, or the position of a substituent, or by altering the

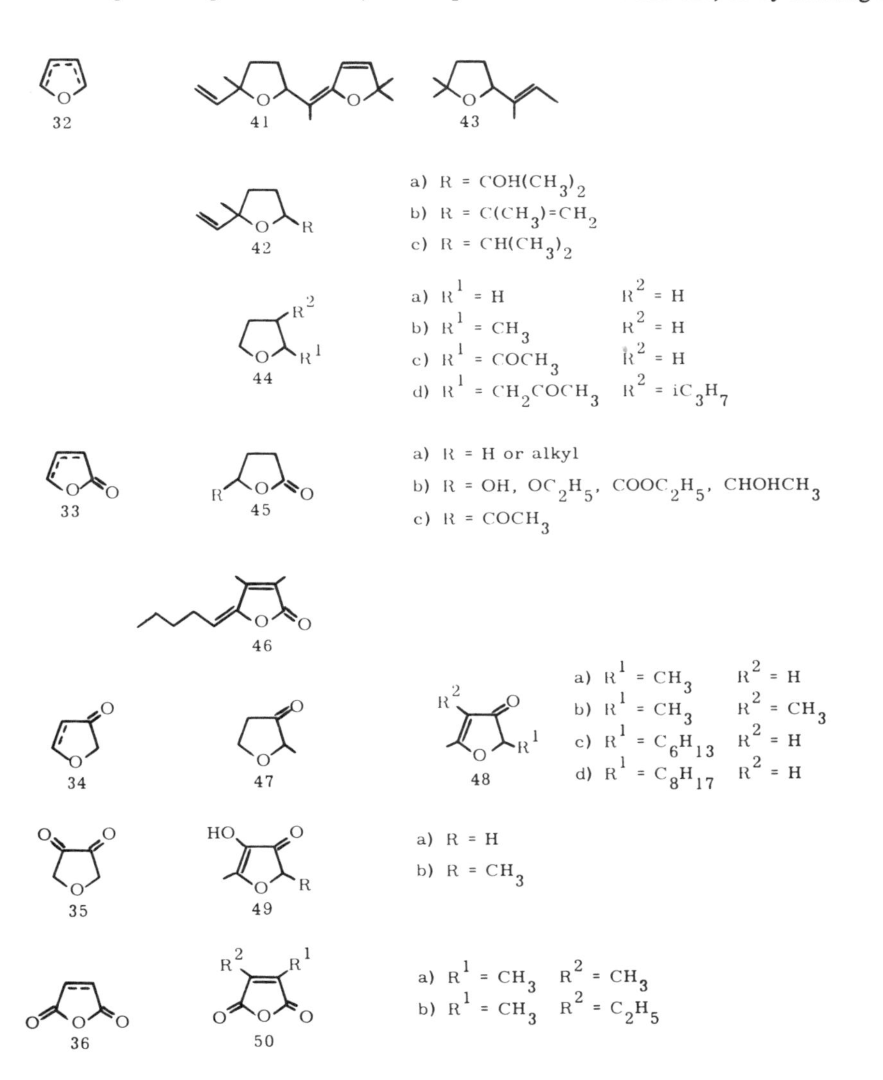

Fig. 2. Monocyclic compounds with mono-oxygenated five-membered rings.

Table 1. Furans: variety of functional substituents.

Chain function	R^5	R^2	Some natural occurrences	Organoleptic properties
Alkyl	–H	$-C_4H_9$	coffee, meat, roasted filberts, potato chips	weak, uncharacteristic
Alkenyl	–H	$-CH{=}CH_2$	roasted filberts, burnt sugar	phenolic, coffee grounds
Alcohol	–H	$-CH_2OH$	coffee, cocoa, fruits, alcoholic beverages, roasted cereals, roasted nuts	sweet, mild odour
Ether	–H	$-CH_2OCH_3$	coffee	strong mustard
Mercaptan	–H	$-CH_2SH$	coffee	strong roasted coffee
Sulfide	$-CH_3$	$-CH_2SCH_3$	coffee, cocoa	mustard, onion-like
Aldehyde	$-CH_3$	–CHO	coffee	burnt, caramel, slightly meaty
Ketone	–H	$-CH_2COCH_3$	coffee, cocoa	rum-like
Diketone	$-CH_3$	$-COCOCH_3$	coffee, rye crispbread	weak, slightly buttery, caramel
Acid	–H	–COOH	raspberry, wine, passion fruit, burnt sugar	weak, uncharacteristic
Ester				
a) furoate	–H	$-COOC_2H_5$	cocoa, passion fruit, roasted peanuts, wine	burnt, buttery, vanilla-like
b) furfuryl ester	–H	$-CH_2OCOC_2H_5$	coffee	pear-like, bitter, nutty
Thioester				
a) thiofuroate	–H	$-COSCH_3$	coffee	cabbage, sulphurous, mercaptan-like
b) furfuryl thioester	–H	$-CH_2SCOCH_3$	coffee	coffee taste
Nitrile	$-CH_3$	–CN	coffee	nutty, bitter almond

oxidation of the ring itself. *Related structures* are thus produced, to be found mainly in the patent literature. This procedure can be carried to a more advanced degree, for instance by permutation or modification of skeleton atoms. Discovery of such substances in a natural flavour is most unlikely.

Classification of heterocyclic structures

Fig. 1 shows the present state of our knowledge and within certain limits the diversity of five and six-membered heterocyclics. It includes 16 monocyclic structures falling into 8 five-membered rings symbolized by (5/X, where X is O, S or N, up to 3 heteroatoms) and 8 six-membered rings (6/X). It also includes 15 condensed bicyclic structures falling into two 5 + 5 ring systems (55/X), nine 5 + 6 ring structures (65/X) and four 6 + 6 ring systems (66/X). The maximum of heteroatoms per skeleton is 3 and that only for compounds containing sulphur. Only in a sulphur-containing skeleton, a trithiolane, does one observe two vicinal heteroatoms, because sulphur has a great tendency to form disulphide or polysulphide bonds (as observed also in aliphatic sulphur compounds).

The limits of Fig. 1 are clearly arbitrary since other heterocyclic rings and systems exist in natural products but they are non-volatile, or of no organoleptic interest or occur only in essential oils rarely used in flavouring. So it would not be very useful to analyse these various structures in detail. Instead we illustrate the heterocyclic skeletons listed with some selected examples. Some classes contain only one known component but others include several dozen products, each with its own organoleptic properties. Choice inevitably involves some omissions.

Selection of some characteristic heterocyclic compounds

Monocyclic structures

5/O Since this class clearly contains the most compounds, it is useful to divide it into subclasses (Fig. 2): the first type includes furans as well as their dihydro and tetrahydro derivatives (Structure 32); the second includes products with one carbonyl group as part of the skeleton: lactones (Structure 33) and 3-furanones (Structure 34). The third includes diketofurans such as 3,4-furandiones (Structure 35) and anhydrides (Structure 36).

Furans are so common in flavours that it would be tedious to mention them all. They are formed by degradation of carbohydrates and are therefore present in nearly all foods. For instance, furans in coffee aroma have been reviewed by Flament et al. (1967). Table 1 illustrates the variety of chemical functions of some furan compounds and indicates their natural occurrence and organoleptic properties.

Fig. 3 illustrates the procedure for filling gaps in a systematic classification for natural aldehydic and ketonic furans. The three missing substances are Structure 37 (Fig. 3: $n = 0$, $R = CH_3$, $R' = C_3H_7$), Structure 38 ($n = 0$, $R = CH_3$, $R' = iC_3H_7$) and Structure 39 ($n = 2$, $R = CH_3$, $R' = H$). They should be identified soon in a natural flavour. Analogous examples of this argument are given by Gautschi et al. (1967) and Gianturco (1967).

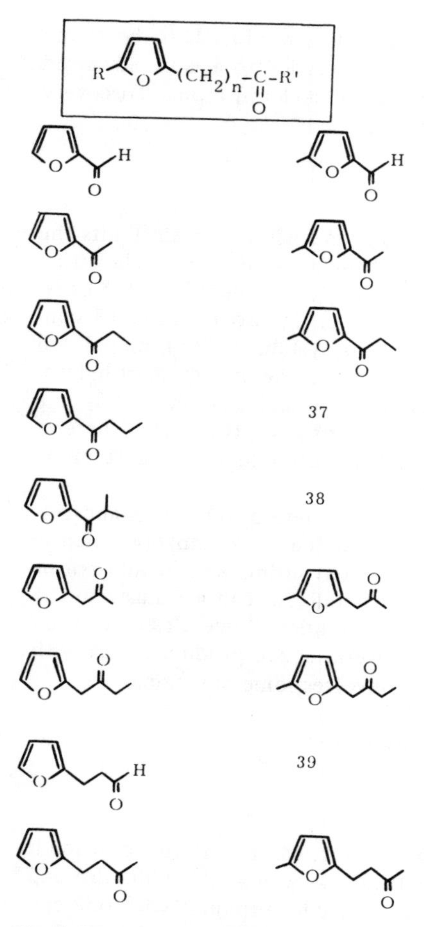

Fig. 3. Example of deduced structures of furan ketones and aldehydes.

In the rare dihydrofurans identified in natural products, the intracyclic double bond is stabilized by conjugation with an exocyclic double bond or by a gem-disubstitution. A rare example of an odorant with such a structure is davana ether (Structure 41, Fig. 2) isolated from *Artemisia pallens* Wall by Thomas & Pitton (1971) and Thomas & Dubini (1974). This sesquiterpenoid also contains a tetrahydrofuran ring. This saturated ring appears in other terpenoid furans such as linalool oxide (Structure 42a) identified in tea, coffee, cocoa, tomato, hop oil, grape, citrus fruits, apricot and passion fruit. Other related structures have been isolated from coffee (Structure 42b), passion fruit (Structure 42c) and citrus fruits (Structure 43). Tetrahydrofuran itself (Structure 44a) and its 2-methyl (Structure 44b) and 2-acetyl (Structure 44c) derivatives, which are more likely formed by pyrolysis, were identified by Stoffelsma et al. (1968) and Friedel et al. (1971) in coffee aroma. 2-Acetonyl-3-

isopropyltetrahydrofuran (Structure 44d) is an unusual component of Burley tobacco flavour (Demole & Berthet, 1972).

The γ-lactones (Structure 33, Fig. 2) are common components of natural flavours. Butanolides commonly have an alkyl group on Position 5 (Structure 45a) but some rings are polysubstituted or possess a functional group (Structure 45b). These γ-lactones are generally identified in fat-containing foods, such as dairy products and meat, and in numerous fruits for whose aroma they are largely responsible. An interesting lactone is solerone (Structure 45c) isolated from wine aroma. It has an "odour typically wine-like, similar to the odour of the cork from a bottle of old premium quality Pinot noir, or as resembling the odour that lingers in a wine glass several hours after an aged wine has been drunk from the glass" (Augustijn et al., 1971).

Butenolides are known with or without a conjugated double bond, but always contain an alkyl or alkylidene chain. Among them, bovolide (Structure 46), identified in butter (Lardelli et al., 1966) and Burley tobacco (Demole & Berthet, 1972) has a celery note explicable by a structure like phthalides identified by Ciamician & Silber (1897) in celery oil.

The 3-furanone ring (Structure 34) exists in food aromas without or with double bond. 2-Methyltetrahydrofuran-3-one (Structure 47) is widely distributed in roasted and manufactured products such as bread, meat, coffee, roasted filberts and

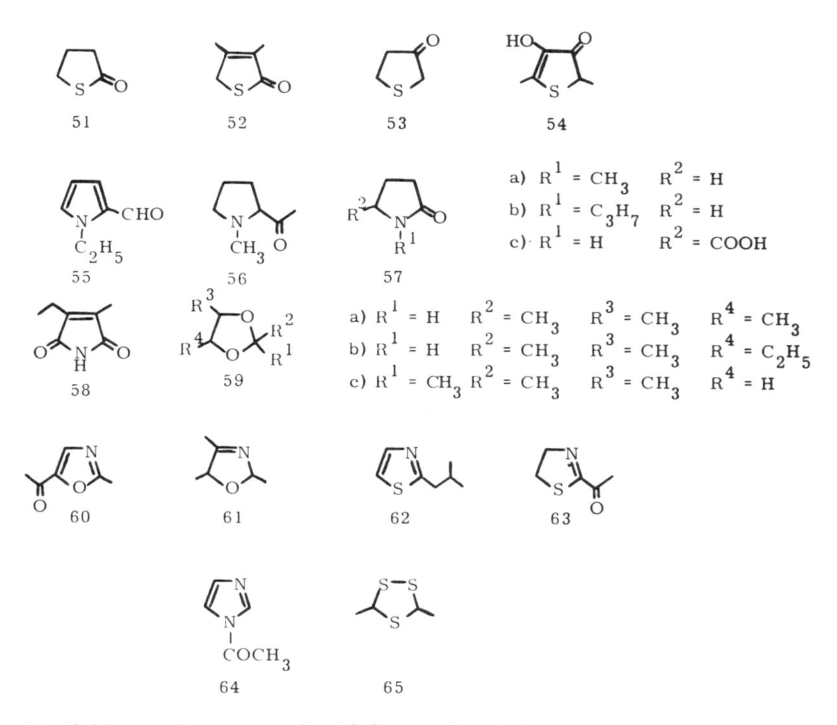

Fig. 4. Monocyclic compounds with five-membered rings.

peanuts, potato chips and rum; it has also been identified in tomato (Buttery et al., 1971). Some alkyl derivatives of *2H*-furan-3-one were isolated from coffee (Structure 48a) by Friedel et al. (1971) and (Structure 49b) by Bondarovich et al. (1967), from white bread (Structure 48a) by Mulders et al. (1972) and from onion flavour (Structure 48c) by Boelens et al. (1971).

The 3,4-furanedione ring (Structure 35) exists essentially in the enolone tautomeric form. Monomethyl (Structure 49a) and dimethyl (Structure 49b) derivatives have a particularly powerful and characteristic note: they were identified in beef broth by Tonsbeek et al. (1968). 2,5-Dimethyl-4-hydroxy-3(*2H*)-furanone (Structure 49b) was also isolated from popcorn, roasted filberts, strawberry and pineapple. Silverstein (1967) characterized its odour as burnt pineapple. Dimethylmaleic (Structure 50a) and methyl ethylmaleic (Structure 50b) anhydrides have been identified in coffee aroma by Stoll et al. (1967).

5/S Thiophene derivatives are not so frequent in flavours as their furan analogs. Aromatic thiophenes are relatively well known as aroma components but only three thiophenones have been identified: thiobutyrolactone (Fig. 4, Structure 51) in coffee by Stoffelsma et al. (1968), a dimethylthiobutenolide (Structure 52) in onion by Boelens et al. (1971) and 4,5-dihydro-3(*2H*)-thiophenone (Structure 53) in roasted products like peanuts, meat and coffee.

2,5-Dimethyl-4-hydroxy-3(*2H*)-thiophenone (Structure 54) is an example of an artificial product with *related structure*, similar to the interesting furanone (Structure 48b) previously mentioned. Compounds of this type were prepared by van den Ouweland and Peer (1970) and by King et al. (1973).

5/N Pyrroles are common in coffee, tea and mushroom, and give these products a burnt and earthy characteristic note. 1-Ethyl-2-formylpyrrole (Structure 55) is a typical component of coffee (Stoll et al., 1967), having a burnt roasted flavour.

Typical of saturated rings is 1-methyl-2-acetyl-pyrrolidine (Structure 56) identified in bread by Hunter et al. (1969).

Pyrrolidinones were isolated from roasted filberts (Structure 57a) (Kinlin et al., 1972), soya bean (Structure 57b) (Manley & Fagerson, 1970) and tomato (Structure 57c) (Bradley, 1960).

Methylethylmaleinimide (Structure 58) was identified in Burley tobacco flavour by Demole & Berthet (1972).

5/OO 1,3-Dioxolanes are not very frequent in food flavours. They are found principally in fruits and wines: 2,4,5-trimethyl-1,3-dioxolane (Structure 59a) in wine, sherry, grape, cranberry and apple; 2,4-dimethyl-5-ethyl-1,3-dioxolane (Structure 59b) in wine and 2,2,4-trimethyl-1,3-dioxolane (Structure 59c) in tomato. These dioxolanes are ketals resulting from condensation of acetaldehyde and acetone with glycols and have an agreeable fruity note.

5/ON This class contains principally oxazoles. Until recently, only one member of this class, 5-acetyl-2-methyloxazole (Structure 60), had been identified in coffee aroma (Stoll et al., 1967), but recent work by Vitzthum and Werkhoff (1974 a,b) identified 20 further alkyloxazoles. Possible pathways for the formation of alkyl and acyl oxazoles are discussed by these authors.

An original and interesting dihydro derivative of this skeleton is 2,4,5-trimethyl-3-oxazoline (Structure 61) identified by Chang et al. (1968) in boiled beef. A patent (Pfizer, 1970) described the synthesis of higher homologues with cocoa, banana, mint, rum and carrot flavours.

5/SN Vitzthum & Werkhoff (1974 a,b) have identified 24 thiazoles in coffee aroma. In general, these alkylthiazoles have green, nutty, roasted and meaty notes. 4-Methyl-5-vinylthiazole identified in cocoa, roasted filberts and passion fruit (Winter & Klöti, 1972) has a fine nutty flavour. 2-Isobutylthiazole (Structure 62) identified in tomato by Viani et al. (1969) has a powerful odour of tomato leaves and various 2-acylthiazoles identified by Stoll et al. (1967) in coffee have a nutty and popcorn flavour.

Tonsbeek et al. (1971) identified an original and interesting dihydro derivative, 2-acetyl-2-thiazoline (Structure 63), in beef broth. It had an intense odour of freshly baked bread crust. Pittet & Hruza (1974) have compared the flavours of thiazole derivatives.

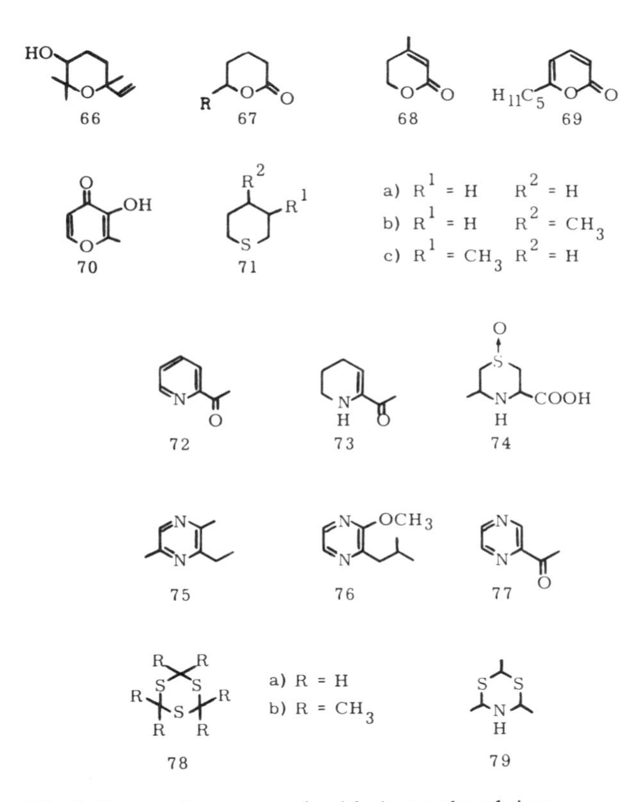

Fig. 5. Monocyclic compounds with six-membered rings.

5/NN Imidazoles are not frequent in volatile food flavours; only 1-acetylimidazole (Structure 64) has been identified in a Burley tobacco flavour (Demole & Berthet, 1972).

5/SSS 2,5-Dimethyl-1,3,4-trithiacyclopentane (trithiolane) (Structure 65) has among others been identified in beef flavour by Chang et al. (1968) and in the mushroom *Boletus edulis* by Thomas (1973). The 2,5-diphenyltrithiolane has recently been identified in guinea-hen weed (*Petiveria alliacea*) by Adesogan (1974).

6/O The compounds of this class (Fig. 5) closely resemble corresponding 5-membered rings (Structure 1, Fig. 1). Terpenoid structures, such as that of the pyran form of linalool oxide (Structure 66) are frequent. The majority of substances in this class are δ-lactones like Structure 67 from beef fats identified by Watanabe & Sato (1968), R = H or alkyl, and Structure 68 isolated from tobacco flavour (Demole & Berthet, 1972), or aromatic products such as 6-pentyl-α-pyrone (Structure 69) isolated by Sevenants & Jennings (1971) from a peach aroma. The most typical compound is maltol (Structure 70), a trace end-product of non-enzymic browning of Maillard type studied by Hodge (1967), who described it as having a fragrant caramel-like aroma, like the burnt sugar of confectionery. Its taste in dilute solutions gives burnt and fruity notes.

6/S Mason et al. (1967) have identified some thiacyclohexanes (Structure 71) in roasted peanuts.

6/N Pyridines and alkylpyridines are widely distributed, mainly in fermented and roasted products (Goldman et al., 1967). In coffee, they are mainly formed by heat degradation of trigonelline (Viani & Horman, 1974). A typical compound, 2-acetylpyridine (Structure 72), has been identified in roasted filberts and peanuts, potato chips, tea and tobacco. One of the most interesting products with this skeleton is 2-acetyl-1,4,5,6-tetrahydropyridine (Structure 73), which has a typical bread aroma (Hunter et al., 1969; Büchi & Wuest, 1971).

6/OO Moshonas & Shaw (1972) identified 1,4-dioxane (Structure 12) in lemon.

6/SN Cycloalliin (3-methyl-1,4-thiazane-5-carboxylic acid *S*-oxide) (Structure 74) is a cyclic sulphoxide amino acid isolated by Virtanen & Matikkala (1959) from onions (*Allium cepa*). A review about it was published by Carson (1967).

6/NN Only recently have the classic alkylpyrazines attracted interest. A recent review by Maga & Sizer (1973) describes the occurrence in foods and the flavour of these substances, which occur principally in heat-treated products: coffee, cocoa, meat, roasted barley, peanuts, pecans, filberts, popcorn, rye crispbread and soya products. Alkyl pyrazines contribute directly to the roasted or cooked flavour of foods: 3-ethyl-2,5-dimethylpyrazine (Structure 75) is one of the most important components in baked potato aroma (Pareles & Chang, 1974). Pyrazines arise by complex interactions between α-amino acids and carbohydrates or small oxycarbon fragments derived from them (Rizzi, 1972). Numerous patents have been devoted to food or tobacco flavouring with alkyl pyrazines. A patent describes the synthesis of the 65 substitution structures of the cycle by alkyl groups containing a maximum of 5 carbon atoms (Flament, 1973).

For 5 years, interest in pyrazines has increased with the almost simultaneous identification of substituted methoxypyrazines in bell peppers (Buttery et al., 1969) and in peas (Murray et al., 1970). 2-Isobutyl-3-methoxypyrazine (Structure 76) has also been isolated from galbanum oil (Bramwell et al., 1969; Burrell et al., 1970). The organoleptic properties of these powerful heterocyclics had been recognized a few years before their identification in a natural flavour (Firmenich & Co., 1965).

Acetylpyrazine (Structure 77) and its homologues were identified in popcorn and roasted peanut (Walradt et al., 1971). Parliment & Epstein (1973) studied the organoleptic properties of pyrazines.

6/SSS Minor et al. (1965) and Wilson et al. (1973) reported the presence of *s*-trithiane (Structure 78a) in chicken flavour, that of trithioacetone (78b) in meat. These products result from the reaction of formaldehyde and acetone with H_2S produced during heating (Wasserman, 1972).

6/SSN Brinkman et al. (1972) identified thialdine (Structure 79) in beef flavour. It is formed by reaction of acetaldehyde with H_2S and NH_3 formed by thermal degradation of animo acids. It has a burnt, fatty and meaty note.

Bicyclic structures

55/OS The sole product of this class is kahweofuran (2-methyl-3-oxa-8-thiabicyclo[3.3.0]-1,4-octadiene) (Structure 80, Fig. 6), identified in coffee by Stoll et al. (1967). The pure substance has a sharp sulphurous odour but traces have a pleasant roasted and smoky note. Proof of the structure and synthesis have been described by Büchi et al. (1971).

55/SS As for kahweofuran, Stoll et al. (1967) first isolated thieno [3,2-*b*]thiophene (Structure 81) from coffee. Mulders (1973) isolated it after a model reaction between ribose and a cysteine-cystine mixture.

65/O Bricout et al. (1967) first identified 2,2,6-trimethyl-7-oxabicyclo[4.3.0.] non-9-en-8-one, (trivial name dihydroactinidiolide) (Structure 82), and oxidation product of β-ionone, in black tea and later in cranberries, currants, tomatoes, crispbread, tobacco and passion fruit. The corresponding 4-ketoderivative has also been identified in tobacco (Demole & Berthet, 1972).

65/O Alkyl and alkylidenephthalides and their dihydro and tetrahydro derivatives are typical components of celery. 3-Butylphthalide (Structure 83) and 3-*n*-butylidenephthalide (*Ligusticum* lactone) (Structure 84) were, for instance, identified in celery and lovage, respectively. For a more detailed bibliography, see A. E. Johnson et al. (1971) and Ohloff (1969).

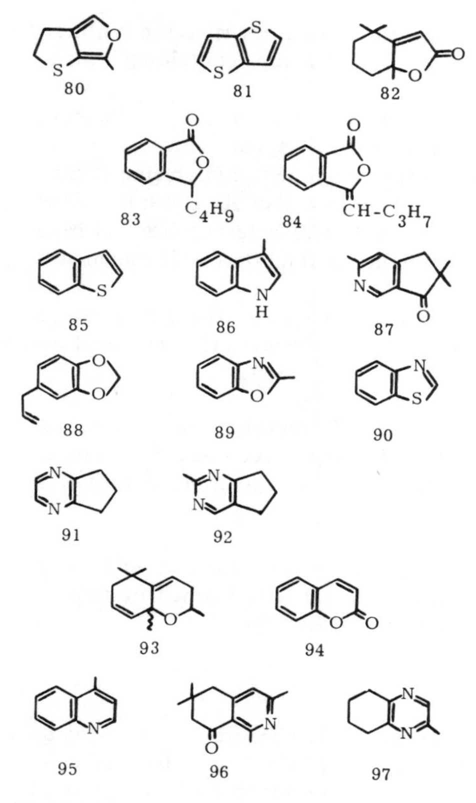

Fig. 6. Bicyclic structures.

65/S Stoll et al. (1967) identified benzo[*b*]thiophene (Structure 85) in coffee and Shipton et al. (1969) identified it in peas.

65/N Various indoles have been identified in food. The most powerful flavouring with this skeleton is skatole (Structure 86), which has been isolated from fish, tea, trassi (shrimp paste), meat and dry milk.

65/N Demole & Demole (1975) isolated from Burley tobacco a terpenoid alkaloid with this unusual skeleton (Structure 87), 3,6,6-trimethyl-5,6-dihydro-7*H*-2-pyrindin-7-one.

65/OO The simplest compound in this class is safrole (1-allyl-3,4-methylenedioxybenzene) (Structure 88) identified in cinnamon, cocoa, mace, pepper, nutmeg and aniseed. Related compounds, principally from spices, have methoxy groups on Position 5 (myristicin), 2,5 (apiole) or 2,3 (dillapiole). This skeleton is also present in piperin, the typical constituent of black pepper (*Piper nigrum*), isolated and identified by Ladenburg in 1894.

65/ON Vitzthum & Werkhoff (1974a,b) identified 2-methylbenzoxazole (Structure 89) and 2,5-dimethylbenzoxazole in coffee flavour.

65/SN Benzothiazole (Structure 90) is a well known constituent of coffee, tobacco, butter, cocoa, cranberries, potatoes, roasted peanuts and dried milk.

65/NN 6,7-Dihydro-5*H*-cyclopenta[*b*]pyrazine (Structure 91) and some of its alkyl derivatives are mentioned for the first time in a patent application (Polak's Frutal Works and Douwe Egberts, 1968). These compounds, which had previously been confused with isomeric alkenylpyrazines, have planty, phenolic, chocolate and peanuts notes. They have been identified in roasted peanuts (B. R. Johnson et al., 1971), green tea (Yamanishi et al., 1973), roasted almonds (Takei et al., 1974) and roasted sesame seed (Manley et al., 1974).

6,7-Dihydro-5*H*-2-methylcyclopenta[*d*]pyrimidine (Structure 92), an artificial substance with a roasted nut and popcorn flavour (Katz et al., 1972) is a distantly related structure in which the position of the heteroatoms has been changed.

66/O One of the most recently identified products of this class is edulan (Structure 93), isolated by Murray et al. (1972) from the juice of the purple passion fruit (*Passiflora edulis* Sims). Both stereoisomers of the substance are present in the fruit. They are described as having a 'rose-like' aroma. Whitfield et al. (1973) elucidated the structure and Adams et al. (1974) synthesized it. Demole & Berthet (1972) identified some compounds with the edulan skeleton in tobacco flavour. This class includes coumarin (Structure 94) and its derivatives which possess important physiological properties. They were identified in cocoa, green tea, whisky, raspberry, cassia and various citrus fruits (Kefford & Chandler, 1970).

66/N Quinoline and its 2-methyl and 6-methyl derivatives are components of whisky and fusel oils. 4-Methylquinoline (lepidine) (Structure 95) has been identified in Burley tobacco (Demole & Berthet, 1972).

66/N 1,3,6,6-Tetramethyl-5,6,7,8-tetrahydroisoquinoline-8-one (Structure 96) is a recently identified component of tobacco (Demole & Demole, 1975).

66/NN Quinoxaline and its 5-methyl derivative are interesting components of roasted foods like coffee (Stoll et al., 1967), roasted peanuts (Walradt et al., 1971) and roasted filberts (Kinlin et al., 1972). Thomas (1973) identified 2-methylquinoxaline in the mushroom *Boletus edulis* and Winter & Klöti (1972) in passion fruit *Passiflora edulis*. 5,6,7,8-Tetrahydroquinoxaline and its 2-methyl derivatives (Structure 97) have been identified in roasted filberts (Kinlin et al., 1972) and in beef flavour (Mussinan et al., 1973).

General considerations

What lesson can be drawn from this review of heterocyclic flavour compounds?
1. With such variety of structures and uneven distribution of numbers in each class, statistical interpretation is difficult. New structures will certainly be discovered, but their number will inevitably diminish when common flavours have been studied and the principal compounds identified. But modern analytical methods will progressively allow the detection of components in amounts even below the sensory threshold, even taking account of synergism. For certain complex aromas, such as coffee, it will be possible to discover new trace components for many years if the analyst has sufficient starting material. Is it still reasonable to investigate these traces of natural products, even when they no longer contribute to organoleptic characteristics? A certain number of such 'ultimate' compounds will probably appear in the analytical literature in the coming years.
2. Model reactions will certainly be more and more used because the products of those reactions may eventually be identified in natural products. Biogenetic rules are applicable only to 'fresh' flavours. Therefore one should examine thoroughly non-enzymic browning reactions, because the majority of foods are consumed after thermal treatment.
3. The distribution of known flavourings over different classes of heterocyclics is meaningless. Research groups often concentrate efforts only on one fraction of the flavour (for instance basic compounds) or on one group of substances. Their results can suddenly augment the number of compounds in a class.
4. Great prudence must be employed when stating organoleptic properties of flavourings. Impurities in a substance synthesized or isolated from the natural mixture can considerably alter the sensory evaluation. It frequently happens that 'savoury' substances of classical or patent literature prove tasteless and odourless after careful purification.

The chemistry of heterocyclic aroma substances is undergoing fruitful development and will continue to provide many rewarding discoveries.

References

Adams, D. R., S. P. Bhatnagar, R. C. Cookson, G. Stanley & F. B. Whitfield, 1974. Synthesis and structures of edulan I and II. J. chem. Soc. D Chem. Commun. 469–470.

Adesogan, E. K., 1974. Trithiolaniacin, a novel trithiolan from *Petiveria alliacea* J. chem. Soc. D Chem. Commun. 906–907.

Augustijn, O. P. H., C. J. van Wijk, C. J. Muller, R. E. Kepner & A. D. Webb, 1971. The structure of solerone, a substituted γ-lactone involved in wine aroma. J. agric. Food Chem. 19: 1128–1130.

Boelens, M., P. J. de Valois, H. J. Wobben & A. van der Gen, 1971. Volatile flavor compounds from onion. J. agric. Food Chem. 19: 984–991.

Bondarovich, H. A., P. Friedel, V. Krampl, J. A. Renner, F. W. Shephard & M. A. Gianturco, 1967. Volatile constituents of coffee. Pyrazines and other compounds. J. agric. Food Chem. 15: 1093–1099.

Bradley, D. B., 1960. The separation of organic and inorganic acid anions in filtered tomato puree by partition chromatography. J. agric. Food Chem. 8: 232–234.

Bramwell, A. F., J. W. K. Burrell & G. Riezebos, 1969. Characterisation of pyrazines in galbanum oil. Tetrahedron Lett. 37: 3215–3216.

Bricout, J., R. Viani, F. Müggler-Chavan, J. P. Marion, D. Reymond & R. H. Egli, 1967. On the

composition of black tea aroma. Helv. chim. Acta 50: 1517-1522.
Brinkman, H. W., H. Copier, J. J. M. de Leuw & S. B. Tjan, 1972. Components contributing to beef flavor. Analysis of the headspace volatiles of beef broth. J. agric. Food Chem. 20: 177-181.
Büchi, G. & H. Wüest, 1971. Synthesis of 2-acetyl-1,4,5,6-tetrahydropyridine, a constituent of bread aroma. J.org. Chem. 36: 609–610.
Büchi, G., P. Degen, F. Gautschi & B. Willhalm, 1971. Structure and synthesis of kahweofuran, a constituent of coffee aroma. J. org. Chem. 36: 199–200.
Burrell, J. W. K., R. A. Lucas, D. M. Michalkiewicz & G. Riezebos, 1970. Characterisation of pyrazines in galbanum oil. Chem. Ind. (London) 1409–1410.
Buttery, R. G., R. M. Seifert, R. E. Lundin, D. G. Guadagni & L. C. Ling, 1969. Characterisation of an important aroma component of bell peppers. Chem. Ind. (London) 490–491.
Buttery, R.G., R. M. Seifert, D. G. Guadagni & L. C. Ling, 1971. Characterisation of additional volatile components of tomato. J. agric. Food Chem. 19: 524–529.
Carson, J. F., 1967. Onion flavor. In: Schultz H. W. (ed.), The chemistry and physiology of flavors, Oregon State University Symposium, AVI Publishing Company, Westport, Connecticut, p. 390–405.
Chang, S. S., C. Hirai, B. R. Reddy, K. O. Herz, A. Kato & G. Sipma, 1968. Isolation and identification of 2,4,5-trimethyl-3-oxazoline and 3,5-dimethyl-1,2,4-trithiolane in the volatile flavour compounds of boiled beef. Chem. Ind. (London) 1639–1641.
Ciamician, G. & P. Silber, 1897. On the constitution of the smelling part of celery oil. Ber. Dtsch. chem. Ges. 30: 1419–1424.
Deck, R. E. & S. S. Chang, 1965. Identification of 2,5-dimethylpyrazine in the volatile flavor compounds of potato chips. Chem. Ind. (London) 1343–1344.
Demole, E. & D. Berthet, 1972. A chemical study of Burley tobacco flavour (*Nicotiana tabacum* L.). 1. Helv. chim. Acta 55: 1866–1882.
Demole, E. & C. Demole, 1975. A chemical study of Burley tobacco flavour (*Nicotiana tabacum* L.). 5. Identification and synthesis of the novel terpenoïd alkaloids 1,3,6,6-tetramethyl-5,6,7,8-tetrahydroisoquinoline-8-one and 3,6,6-trimethyl-5,6-dihydro-7H-2-pyrindin-7-one. Helv. chim. Acta, in press.
Firmenich & Co. (Patentee), 1965. Methoxy- and methylthiopyrazines. Fr. Pat. 1. 391.212.
Flament, I., 1973. Aromatic compositions. Swiss Pat. 540.016.
Flament, I., F. Gautschi, M. Winter, B. Willhalm & M. Stoll, 1967. The furanic components of coffee aroma: some chemical and spectroscopical aspects. In: 3rd International Colloquium on Coffee Chemistry, Trieste, p. 197–215.
Friedel, P., V. Krampl, T. Radford, J. A. Renner, F. W. Shephard & M. A. Gianturco, 1971. Some constituents of the aroma complex of coffee. J. agric. Food Chem. 19: 530–532.
Gautschi, F., M. Winter, I. Flament, B. Willhalm & M. Stoll, 1967. The chemistry of coffee aroma. A survey of present knowledge. 3rd International Colloquium on Coffee Chemistry, Trieste, p. 67–76.
Gianturco, M. A., 1967. Coffee flavor. In: Schultz H. W. (ed.), The chemistry and physiology of flavors, Oregon State University symposium, AVI Publishing Company, Westport, Connecticut, p. 431–449.
Goldman, I. M., J. Seibl, I. Flament, F. Gautschi, M. Winter, B. Willhalm & M. Stoll, 1967. On coffee aroma. 2. Pyrazines and pyridines. Helv. chim. Acta 50: 694–705.
Hodge, J. E., 1967. Origin of flavor in foods: nonenzymic browning reactions. In: Schultz H. W. (ed.), The chemistry and physiology of flavors, Oregon State University symposium, AVI Publishing Company, Westport, Connecticut, p. 465–491.
Hunter, I. R., M. K. Walden, J. R. Scherer & R. E. Lundin, 1969. Preparation and properties of 1,4,5,6-tetrahydro-2-acetopyridine, a cracker-odor constituent of bread aroma. Cereal Chem. 46: 189–195.
Johnson, A. E., H. E. Nursten & A. A. Williams, 1971. Vegetable volatiles: a survey of components identified. 2. Chem. Ind. (London) 1212–1224.
Johnson, B. R., G. R. Waller & A. L. Burlingame, 1971. Volatile components of roasted peanuts: basic fraction. J. agric. Food Chem. 19: 1020–1024.
Katz, I., R. A. Wilson, W. J. Evers, M. H. Vock, G. W. Verhoeven & J. Sieczkowski (IFF), 1972. Bicyclic pyrimidine derivatives, preparation and application. West German Patent, 2.141.916.

Kefford, J. F. & B. V. Chandler, 1970. The chemical constituents of citrus fruits. In: Advances in Food Research, supplement 2, Academic Press, p. 106–111.
King, B., E. Demole & A. F. Thomas, 1973. Procedure for the preparation of heterocyclic derivatives. Swiss Pat. 539.631.
Kinlin, T. E., R. Muralidhara,A. O. Pittet, A. Sanderson & J. P. Walradt, 1972. Volatile components of roasted filberts. J. agric. Food Chem. 20: 1021–1028.
Ladenburg, A. & M. Scholtz, 1894. Synthese der Piperinsaüre und des Piperins. Ber. Dtsch. chem. Ges. 27: 2958–2960.
Lardelli, G., G. Dijkstra, P. D. Harkes & J. Boldingh, 1966. New γ-lactone found in butter. Rec. Trav. Chim. 85: 43–55.
Maga, J. A. & C. E. Sizer, 1973. Pyrazines in foods: a review. J. agric. Food Chem. 21: 22-30.
Manley, C. H. & T. S. Fagerson, 1970. Major volatile neutral and acid compounds of hydrolyzed soy protein. J. Food Sci. 35: 286–291.
Manley, C. H., P. P. Vallon & R. E. Erickson, 1974. Some aroma components of roasted sesame seed (*Sesamum indicum* L.). J. Food Sci. 39: 73–76.
Mason, M. E., B. Johnson & M. C. Hamming, 1967. Volatile components of roasted peanuts. The major monocarbonyls and some noncarbonyl components. J. agric. Food Chem. 15: 66–73.
Minor, L. J., A. M. Pearson, L. E. Dawson & B. S. Schweigert, 1965. Chicken flavor: the identification of some chemical components and the importance of sulfur compounds in the cooked volatile fraction. J.Food Sci. 30: 686-696.
Moshonas, M. G. & P. E. Shaw, 1972. Analyses of flavor constituents from lemon and lime essence. J. agric. Food Chem. 20: 1029–1030.
Mulders, E. J., 1973. Volatile components from the non-enzymic browning reaction of the cysteine/cystine-ribose system. Z. Lebensm. Unters. Forsch. 152: 193-201.
Mulders, E. J., H. Maarse & C. Weurman, 1972. The odour of white bread. 1. Analysis of volatile constituents in the vapour and aqueous extracts. Z. Lebensm. Unters. Forsch. 150: 68–74.
Murray, K. E., J. Shipton & F. B. Whitfield, 1970. 2-Methoxy-pyrazines and the flavour of green peas (*Pisum sativum*). Chem. Ind. (London) 897–898.
Murray, K. E., J. Shipton & F. B. Whitfield, 1972. Volatile constituents of passionfruit, *Passiflora edulis*. Aust. J. Chem. 25: 1921–1933.
Mussinan, C. J., R. A. Wilson & I. Katz, 1973. Isolation and identification of pyrazines present in pressure-cooked beef. J. agric. Food Chem. 21: 871–872.
Ohloff, G., 1969. Chemistry of odoriferous and tasting compounds. In: Fortschritte der chemischen Forschung, Band 12, Heft 2, p. 185–251.
Ouweland, G. A. M. van den & H. G. Peer, 1970. Aromatic substances and methods of preparation. West German Patent 1.932.800.
Pareles, S. R. & S. S. Chang, 1974. Identification of compounds responsible for baked potato flavor. J.agric. Food Chem. 22: 339–340.
Parliment, T. H. & M. F. Epstein, 1973. Organoleptic properties of some alkyl-substituted alkoxy and alkylthiopyrazines. J. agric. Food Chem. 21: 714–716.
Pfizer Inc., 1970. 2,5-Dialkyl-2-oxazolines. U.S. Pat. 3.769.293.
Pittet, A. O. & D. E. Hruza, 1974. Comparative study of flavor properties of thiazole derivatives. J. agric. Food Chem. 22: 264–269.
Polak's Frutal Works – Douwe Egberts, 1968. Taste and flavoring substances. Dutch patent application 68.12899.
Reichstein, T. & H. Staudinger, 1926. Brit Pat. 246.454 & 260.960.
Rizzi, G. P., 1972. A mechanistic study of alkylpyrazine formation in model systems. J. agric. Food Chem. 20: 1081–1085.
Schrötter, H., 1879. On a basic constituent of fusel oil. Ber. Dtsch. chem. Ges. 12: 1431–1432.
Sevenants, M. R. & W. G. Jennings, 1971. Occurrence of 6-pentyl-α-pyrone in peach essence. J. Food Sci. 36: 536.
Shipton, J., F. B. Whitfield & J. H. Last, 1969. Extraction of volatile compounds from green peas (*Pisum sativum*). J. agric. Food Chem. 17: 1113–1118.
Silverstein, R. M., 1967. Pineapple flavor. In: Schultz H. W. (ed.), The chemistry and physiology of flavors, Oregon State University symposium, AVI Publishing Company, Westport, Con-

necticut, p. 450–461.
Stoffelsma, J., G. Sipma, D. K. Kettenes & J. Pijpker, 1968. New volatile components of roasted coffee. J. agric. Food Chem. 16: 1000–1004.
Stoll, M., M. Winter, F. Gautschi, I. Flament & B. Willhalm, 1967. On coffee aroma. 1. Helv. chim. Acta 50: 628–694.
Takei, Y., K. Shimada, S. Watanabe & T. Yamanishi, 1974. Volatile components of roasted almonds: basic fraction. Agr. biol. Chem. 38: 645–648.
Thomas, A. F., 1973. An analysis of the flavour of the dried mushroom, *Boletus edulis*. J. agric. Food Chem. 21: 955–958.
Thomas, A. F. & G. Pitton, 1971. The isolation, structure and synthesis of davana ether, an odoriferous compound of the oil of *Artemisia pallens* Wall. Helv. chim. Acta 54: 1890–1891.
Thomas, A. F. & R. Dubini, 1974. Terpenoids derived from linalyl oxide. 4. The oxidation of davanone. Isolation and synthesis of the davana ethers, sesqui-terpenes of *Artemisia pallens*. Helv. chim. Acta 57: 2076–2081.
Tonsbeek, C. H. T., A. J. Plancken & T. van den Weerdhof, 1968. Components contributing to beef flavor. Isolation of 4-hydroxy-5-methyl-3(*2H*)-furanone and its 2,5-dimethyl homolog from beef broth. J. agric. Food Chem. 16: 1016–1021.
Tonsbeek, C. J. T., H. Copier & A. J. Plancken, 1971. Components contributing to beef flavor. Isolation of 2-acetyl-2-thiazoline from beef broth. J.agric. Food Chem. 19: 1014–1016.
Viani, R. & I. Horman, 1974. Thermal behavior of trigonelline. J. Food Sci. 39: 1216–1217.
Viani, R., J. Bricout, J. P. Marion, F. Müggler-Chavan, D. Reymond & R. H. Egli, 1969. On the composition of tomato aroma. Helv. chim. Acta 52: 887–891.
Virtanen, A. I. & E. T. Matikkala, 1959. The structure and synthesis of cycloalliin isolated from *Allium cepa*. Acta chem. Scand. 13: 623–626.
Vitzthum, O. G. & P. Werkhoff, 1974a. Newly discovered nitrogen containing heterocycles in coffee aroma. Z. Lebensm. Unters. Forsch. 156: 300–307.
Vitzthum, O. G. & P. Werkhoff, 1974b. Oxazoles and thiazoles in coffee aroma. J. Food Sci. 39: 1210–1215.
Walradt, J. P., A. O. Pittet, T. E. Kinlin, R. Muralidhara & A. Sanderson, 1971. Volatile components of roasted peanuts. J. agric. Food Chem. 19: 972–979.
Wasserman, A. E., 1972. Thermally produced flavor components in the aroma of meat and poultry. J. agric. Food Chem. 20: 737–741.
Watanabe, K. & Y. Sato, 1968. Gas chromatographic identification of aliphatic γ- and δ-lactones obtained from beef fats. Agric. biol. Chem. 32: 191–196.
Whitfield, F. B., G. Stanley & K. E. Murray, 1973. Concerning the structures of edulan I and II. Tetrahedron Lett. 95–98.
Wilson, R. A., C. J. Mussinan, I. Katz & A. Sanderson, 1973. Isolation and identification of some sulfur chemicals present in pressure-cooked beef. J. agric. Food Chem. 21: 873–876.
Winter, M. & R. Klöti, 1972. On the aroma of yellow passion fruit (*Passiflora edulis* f. *flavicarpa*). Helv. chim. Acta 55: 1916–1921.
Yamanishi, T., S. Shimojo, M. Ukita, T. Kawashima & Y. Nakatani, 1973. Aroma of roasted green tea (Hoji-cha). Agr. biol. Chem. 37: 2147–2153.

Proc. int. Symp. Aroma Research, Zeist, 1975. Pudoc, Wageningen.

Aromatization and international legislation[1]

Short communication

H. van den Dool

International Flavors & Fragrances (Europe), Liebergerweg 72-98, Hilversum, the Netherlands

It is to be regretted very much that the paper on 'International legislative control of flavouring substances in food' has not become available for study before the symposium. It is even more regrettable that the author, who has been announced as belonging to the United Kingdom's Ministry of Agriculture, Fisheries and Food, is not present here for discussion. It would have been interesting to have heard from him whether his paper represents his personal opinion or whether it reflects the British official point of view. This is of interest because the paper presents a somewhat biased view on the regulations for flavourings since it discusses only a positive permitted list system and completely neglects the existence of other systems, such as the 'mixed' system in use in Spain, the German Federal Republic, Italy and in the near future also in the Netherlands and in Finland.

Types of regulation

There is no doubt that everywhere the protection of public health and of certain consumer's interests is the principle goal of the food law. In the application of the general provisions of such a law, one can distinguish between vertical regulations for commodities such as bread, soft drinks and milk, and horizontal regulations for food additives such as colouring materials, preserving agents and emulsifying agents. The questions arising are:

a. which system will be chosen for legislative control of food additives

b. whether flavouring agents can be dealt with in the same way as other classes of food additives.

To start with the latter question, it is clear that flavouring agents differ from these other classes for the following reasons, which are partially discussed also in Dr Coomes' paper.

1. Most foods already contain flavouring substances, whether these are already present in the food's raw state or whether they develop during the preparation of the food. To food flavours proper, one must add the flavouring agents intentionally used for improvement of the palatability of the food and by this way in many cases improving the physiological efficiency.

1. Reaction to the paper of Dr Coomes, 28 May, 1975.

2. Contrary to other classes of food additives the number of flavouring compounds is already enormous and increasing daily by detection of new food flavouring compounds proper.
3. The use level of each flavouring compound in the finished food is mostly very low.
4. Contrary to most other food additives such as preservatives and emulsifying agents, the consumer becomes easily aware of added flavourings in the foods offered for consumption.

This distinct position of the flavouring compounds ought to be reflected in a different legislative treatment. Every other class of food additives, such as antioxidants, colouring agents, preservatives and emulsifying agents due to the low number used in food technology and their almost all being practically not food-identical may be regulated by positive lists. For flavouring additives for the reasons stated, another system has to be chosen, such as the mixed system, in which some flavouring agents are on a positive list, others on a negative list.

Priority in toxicological research

Independent of the system to be used the only criterion to be applied in assessing the admissibility of a flavouring agent clearly is its degree of toxicity under the conditions of use. However, to assess this toxicity for the many thousands of flavouring compounds would place an enormous burden on the toxicological investigation capacity available. Expansion of this capacity is not justified by the gain, if any, in protection of public health. To make the most economic use of the capacity available, priorities have to be established. A non-priority criterion is found in the human experience (mostly indicated by long history of use) with a food. Foods consumed already over a long period of time without noticeable adverse effects will have a very low probability of being possibly toxic. The same holds, of course, for the flavouring components of these foods at the normal level of use. At the same time the important number of these 'nature-identicals' (food-identicals would have been more appropriate) present at low levels means that their metabolism in the human body is spread over different pathways instead of overloading perhaps a single one.

What has been said here is one of the principal reasons that in the mixed system naturals and nature-identicals are not listed. It is however obvious that in food compounds may occur which at some concentration may be noxious. Examples are safrole, thujone, benzpyrene, etc. These compounds are listed, together with limits for their use, on negative or restrictive lists. It may be worthwile to note that the classification in naturals, nature-identicals and artificials has been accepted by the Codex Alimentarius Committee on Food Additives also.

This brings up the category of artificial flavouring compounds, being those that have not been identified so far in food. These compounds – about 200 to 300 are in use today – are readily defined, also for purity. Many of these lack the long history of use, at least in comparison to the flavourings in naturals. Obviously they should have priority in toxicological investigation.

It seems useful to point out that the term 'artificial' is used here in the restricted sense of 'not being identified in food'. The word 'synthetic' comprises as well artificials as 'nature-identicals'. Confusion should be avoided by proper use of these terms. In the mixed system the artificials, together with limits for use, figure on a positive list.

Abuse of the lists of the Council of Europe's Subcommittee

Before discussing a second point of great importance with regard to the choice of legislative system. I should like to point to a common fallacy about the Council of Europe's study. This study is not meant to be a model for a legislation with permitted lists as suggested also in the paper of Dr Coomes. It is merely an exercise in toxicology and can serve as a guidance in toxicological matters and should be used as such only. Every other use amounts to abuse. To illustrate this it might be pointed out that ethanol and *D*-glucose still figure on the Council of Europe's list of 'artificial' flavouring substances not fully evaluated. In the preambule of the list it is stated: "Inclusion of a substance in this list means that it is not possible to recommend its use in food on the basis of information presently available to the experts of the Working Group."

Difficulties in quantitative analysis

The other point of principal interest is the old adagium that a law is only as good as its enforcement is possible. As an analytical chemist, and I am sure many of you will have the same experience, I know how difficult it is to isolate quantitatively a flavouring from a food. A difficulty which readily is understandable if we remember yesterday's lectures on technical subjects. The flavouring isolated usually will consist of a mixture of naturals, e.g. essential oils, their fractions of vegetable extracts, with nature-identicals and possibly a few artificials.

Flavour formulae containing 50 or more ingredients are no exception and remembering that some of the natural ingredients may show in a chromatogram hundreds of peaks it is easily understood that analytical check of such an isolated flavouring is not an easy task, the more so as the food flavouring compounds proper will have to be included in the picture.

It is clear that under a positive-list system every peak will have to be identified properly, for instance by gas chromatographic retention and by mass spectrum. In a mixed system the analyst only has to look for those on the restrictive list and for artificials.

Within a country the difficulty of analytical checking can be replaced by factory inspection. However, imported foods should also be checked, otherwise there will arise discrimination against national manufacturers.

Conclusion

From what has been said, it is obvious that for practical reasons the mixed system is preferred. It protects adequately both the consumer and the bonafide manufacturer.

The advantages of the positive, permitted list system as stated by Dr Coomes would result in the authorities assuming the responsibility for the safety-in-use of a flavouring additive. There will be much doubt whether authorities are inclined to accept this legal responsibility. If they would so, this would be only on base of expensive, in money, time and animals, toxicological investigations to be carried out by the flavour manufacturer. With a view to the restricted toxicological capacity available this would create problems. Moreover early publications would deprive the flavour manufacturer from what he feels is his industrial property. The net result of all this would be that practically an end would come to all kind of flavour research. A result that would also have its impact on the consumer's way of life and, to say the least, would make life very dull.

In the short time available for an intervention it was not possible to discuss many points in greater detail. I hope I have been able to show you that legislation in the flavour field has more aspects and possibilities than the one discussed in Dr Coomes paper.

Proc. int. Symp. Aroma Research, Zeist, 1975. Pudoc, Wageningen.

Wide-bore glass capillary columns in gas chromatography of aroma components

Short communication

H. T. Badings

Netherlands Institute for Dairy Research (NIZO), P. O. Box 20, Ede, the Netherlands

Capillary columns made of glass are preferable for gas chromatography of polar or labile compounds, because they have a lower catalytic activity and adsorptivity. Wide-bore glass capillary columns have some advantages over the narrow-bore types as they can easily be installed in gas-chromatographic instruments for packed columns, without drastic modifications of the detector and injector. They allow also analysis of larger samples, the collection of trace fractions and the sensory evaluation of emerging peaks, simultaneously with chromatographic recording. For this purpose, an all-glass splitter, glass-capillary collection traps and a device for direct reinjection of fractions from these traps, were designed.

Wide-bore glass capillary columns were prepared, coated with different polar or apolar materials. They had to be properly etched, carbonized, deactivated, and coated. We now use the prepared capillaries in routine aroma research and pollution studies.

List of participants

M. Apetz (LM-Chem.), Institut für Chemisch-Technische Analyse, 1 Berlin 65, Seestrasse 13, Bundesrepublik Deutschland.

Dr Ir H. T. Badings, Netherlands Institute for Dairy Research (NIZO), P.O. Box 20, Ede, the Netherlands.

Drs R. Belz, CIVO-TNO, Utrechtseweg 48, Zeist, the Netherlands.

Ir J. M. H. Bemelmans, CIVO-TNO, Utrechtseweg 48, Zeist, the Netherlands.

H. Boelens, Naarden International N.V., P. O. Box 2, Naarden/Bussum, the Netherlands.

H. Brouwer, Polak's Frutal Works B.V., Nijverheidsweg 7 Zd., P.O. Box 3, Amersfoort, the Netherlands.

Dr T. J. Coomes (not present), Ministry of Agriculture Fisheries and Food, Great Westminster House, Horseferry Road, London S.W.1, England.

J. H. Dhont, CIVO-TNO, Utrechtseweg 48, Zeist, the Netherlands.

Dr H. van den Dool, International Flavors & Fragrances (Europe), Liebergerweg 72-98, Hilversum, the Netherlands.

Prof. Dr F. Drawert, Institut für Chemisch-Technische Analyse und Chemische Lebensmitteltechnologie der Technischen Universität München, 8050 Freising-Weihenstephan, Bundesrepublik Deutschland.

Dr K.-H. Fischer, Deutsche Forschungsanstalt für Lebensmittelchemie, 8 München 40, Leopoldstrasse 175, Bundesrepublik Deutschland.

Dr I. Flament, Firmenich SA, Research Division, P.O. Box 239, CH-1211 Geneva 8, Switzerland.

L. J. van Gemert, CIVO-TNO, Utrechtseweg 48, Zeist, the Netherlands.

Dr J. W. Gramshaw, Procter Dept of Food and Leather Science, University of Leeds, Leeds LS2 9JT, England.

Drs P. J. Groenen, CIVO-TNO, Utrechtseweg 48, Zeist, the Netherlands.

Dr W. Grosch, Deutsche Forschungsanstalt für Lebensmittelchemie, 8 München 40, Leopoldstrasse 175, Bundesrepublik Deutschland.

Dr H. L. Hansen, The Royal Danish School of Educational Studies, Dept of Nutrition and Biochemistry, Emdrupvej 101, DK-2400 Copenhagen NV, Denmark.

Dr H. G. Haring, Naarden International N.V., P. O. Box 2, Naarden/Bussum, the Netherlands.

Dr J. G. de Heus, D.E.J. International Research Co. B.V., Keulsekade 143, Utrecht, the Netherlands.

H. P. Kallio (M.A.), Technical Research Centre of Finland, Lab. for Food Research and Technology, Dept of Biochemistry, University of Turku, S.F. 20500 Turku 50, Finland.

Ir P.J.A.M. Kerkhof, Eindhoven University of Technology Dept of Chemical Engineering, Lab. for physical technology, Insulindelaan, P.O. Box 513, Eindhoven, the Netherlands.

Dr E. P. Köster, Psychological Laboratory, University of Utrecht, Varkenmarkt 2, Utrecht, the Netherlands.

Drs P. J. Kühn, Naarden International N.V., P. O. Box 2, Naarden/Bussum, the Netherlands.

Dr D. G. Land, Agricultural Research Council, Food Research Institute, Colney Lane, Norwich NR4 7UA, England.

Prof. Dr R. R. Linko, Dept of Biochemistry, University of Turku, S.F. 20500 Turku 50, Finland.

Dr H. Maarse, CIVO-TNO, Utrechtseweg 48, Zeist, the Netherlands.

Prof. Dr H. G. Maier, Institut für Lebensmittelchemie der Technischen Universität Braunschweig, D-33 Braunschweig, Fasanenstrasse 3, Bundesrepublik Deutschland.

Ms A. Matteï, Lab. de Physiologie Végétale Appliquée, Université Paris VI, 4, Place Jussieu, Tour 53, 75230 Paris Cedex 05, France.

Dr R. Neeter, Netherlands Institute for Dairy Research (NIZO), P.O. Box 20, Ede, the Netherlands.

Ir M. C. ten Noever de Brauw, CIVO-TNO, Utrechtseweg 48, Zeist, the Netherlands.

Dr H. E. Nursten, Procter Dept of Food and Leather Science, University of Leeds, Leeds LS2 9JT, England.

Dr L. Nykänen, Research Laboratories of the State Alcohol Monopoly (Alko), P.O. Box 350, 00101 Helsinki 10, Finland.

Ms N. Paillard, Lab. de Physiologie Végétable Appliquée, Université Paris VI, 4, Place Jussieu, Tour 53, 75230 Paris Cedex 05, France.
Dr R. L. S. Patterson, Agricultural Research Council, Meat Research Institute, Langford, Bristol BS18 7DY, England.
Dr H. G. Peer, Unilever Research Duiven, P.O. Box 7, Zevenaar, the Netherlands.
Prof. Dr W. Pilnik, Landbouwhogeschool, Afdeling Levensmiddelentechnologie, De Dreyen 2, Wageningen, the Netherlands.
Dr A. J. Prinsen, Keuringsdienst van Waren, Keizersgracht 732-734, Amsterdam, the Netherlands.
Dr W. Renold, Firmenich SA, Research Division, P.O. Box 239, CH-1211 Geneva 8, Switzerland.
Prof. Dr D. Reymond, Nestlé Products Technical Assistance Co. Ltd, Ch-1814, La Tour de Peilz, Switzerland.
Dr F. Rijkens, Naarden International N.V., P.O. Box 2, Naarden/Bussum, the Netherlands.
Ir J. van Roekel, D.E.J. International Research Co. B.V., Keulsekade 143, Utrecht, the Netherlands.
Dr M. Rothe, Zentralinstitut für Ernährung Potsdam-Rehbrücke der Akademie der Wissenschaften der DDR, 1505 Bergholz-Rehbrücke, Arthur-Scheunert-Allee 114-116, DDR.
C. P. Rovers, Food Industries B.V. (Unilever N.V.), Maarssenbroeksedijk 2, P.O. Box 10, Maarssen, the Netherlands.
Dr H. Russwurm Jr, Norwegian Food Research Institute, P.O. Box 50, 1432-Ås-N.L.H., Norway.
Ms P. Salo (Ph.L.), Research Laboratories of the State Alcohol Monopoly (Alko), P.O. Box 350, 00101 Helsinki 10, Finland.
Drs J. Schaefer, CIVO-TNO, Utrechtseweg 48, Zeist, the Netherlands.
Dr P. Schreier, Institut für Chemisch-Technische Analyse und Chemische Lebensmitteltechnologie der Technischen Universität München, D-8050 Freising-Weihenstephan, Bundesrepublik Deutschland.
Prof. Dr J. Solms, Swiss Federal Institute of Technology, Dept of Food Science, 2. Universitätstrasse, CH-8006 Zürich, Switzerland.
Drs S. van Straten, CIVO-TNO, Utrechtseweg 48, Zeist, the Netherlands.
Drs J. Strating, Heineken Brewery, c/o Utrechtseweg 48, Zeist, the Netherlands.
Prof. Dr E. von Sydow, Swedish Institute for Food Preservation Research (S.I.K.), Fack, S 400 21 Göteborg 16, Sweden.
Prof. Dr H. A. C. Thijssen, D.E.J. International Research, B.V., Keulsekade 143, Utrecht, the Netherlands.
Drs R. Timmer, Naarden International N.V., P.O. Box 2, Naarden/Bussum, the Netherlands.
Prof. Dr R. Tressl, Institut für Chemisch-Technische Analyse der Technischen Universität Berlin, 1 Berlin 65, Seestrasse 13, Bundesrepublik Deutschland.
Dr P. Werkhoff, Hag Aktiengesellschaft Bremen, Wissenschaftliche Abteilung, 28 Bremen, Hagstrasse, Bundesrepublik Deutschland.
Dr A. A. Williams, University of Bristol, Department of Agriculture and Horticulture Research Station, Long Ashton, Bristol BS18 9AF, England.